D1207448

Wireless Technician's Handbook

Second Edition

For a complete listing of the *Artech House Mobile Communications Series*,
turn to the back of this book.

Wireless Technician's Handbook

Second Edition

Andrew Miceli

Artech House
Boston • London
www.artechhouse.com

Library of Congress Cataloging-in-Publication Data
Miceli, Andrew.
Wireless technician's handbook/Andrew Miceli.—2nd ed.
p. cm. — (Artech House mobile communications series)
Includes bibliographical references and index.
ISBN 1-58053-357-4 (alk. paper)
1. Wireless communication systems. I. Title II. Series.

TK5103.2.M53 2003
621.382—dc21 2003052297

British Library Cataloguing in Publication Data
Miceli, Andrew
Wireless technician's handbook. —2nd ed.—
(Artech House mobile communications series)
1. Wireless communication systems I. Title
621.3' 82

ISBN 1-58053-357-4

Cover design by Igor Valdman

International Standard Book Number: 1-58053-357-4
Library of Congress Catalog Card Number: 2003052297

10 9 8 7 6 5 4 3 2 1

Contents

Foreword

I have noticed over my 20-year career in wireless engineering that the most effective technicians and engineers are those who understood the basic fundamentals of radio frequency engineering. Too often, technicians understand how to follow the test procedures or troubleshooting guidelines to make the RF measurements, but does not understand what they are measuring and why. It is important to understand the answers to the what and why questions, since often the unexpected happens on site and adjustments have to made on the fly. This is impossible when the technician does not have the proper training or understanding of the fundamentals.

Due to the current state of the telecommunications industry, it is not always possible to get the proper training since budgets are tight. The responsibility often falls on individuals to make up for this lack of training on their own in order to meet the requirements of their jobs. Furthermore, field engineers and technicians find it frustrating that many of the available wireless textbooks are very academic with minimum hands-on experience. I feel that this book offers the opportunity to gain much of the basic technical knowledge necessary to have a long and successful career in the wireless telecommunications field.

This updated edition lays the foundation of wireless engineering by covering the first generation analog cellular systems (AMPS) and the second generation digital cellular systems (GSM, TDMA, and CDMA) before moving into the new 2.5G (GPRS/EDGE and CDMA2000 1X) and third generation or 3G wireless technologies (CDMA2000, 1xEV-DO, and UMTS) that deliver increased voice capacity and/or high-speed packet data services.

Ken Geisheimer
Senior Manager, CDMA Access Product Line Management, Nortel Networks
Richardson, Texas
July 2003

Acknowledgments

Amazingly, this update to my original text seems to have taken more energy and time than the original. I must first thank the staff of my publisher for their patience on this project.

Obviously, a project like this is not even remotely possible without help from expert colleagues and good friends, and even less possible without a supportive and understanding family. Fortunately, I have both. My wife Stephanie, sons Joel, Mitchell, and Daniel, and daughter Cassandra put up with my late hours and always supported my efforts.

For this update, several colleagues were especially important in providing me with help and technical information. Among them are Amitaabh Malhotra, Melissa Summers, Tom Jones, Mike Barrick, and Markku Salkola. I also must thank my good friends in the Alltel RF group, from whom I have learned a tremendous amount over the last few years, including Amir Shahegh, Mike Stafford, Tory Cruse, Chuck Gee, Billy Spann, and Ron Krueger. Last, but not least, I must thank the management at Superconductor Technologies that allowed me to take on this project, with special thanks to Rich Conlon, Bob Johnson, Peter Thomas, and a good friend and mentor, Ray Cotten.

Of course, this new update is only an update to the original version. Thus, I must again thank those who contributed to the original text: Allan Mowat, Christopher Rix, Guenther Klenner, Achim Grolman, Allen Shuff, Russ Byrd, and Cyrille Damany.

CHAPTER 1

Reviewing the Basics

1.1 A Review for the Technician, an Introduction for the Newcomer

In order to understand the complexities of today's digital wireless formats, it is essential that technicians and engineers have a very sound foundation in basic *radio frequency* (RF) and digital principles. The technicians who completely understand these fundamentals have a tremendous advantage over their peers. This chapter will explain many of the common concepts of RF as well as review some commonly used digital modulation techniques.

1.2 Domains: Time and Frequency

A word used in the electronic world quite often is *domain*. If we look at a basic XY chart, the X component is called the domain (see Figure 1.1). Essentially, it is the basis for the graphical representation of an item—in our case, an RF signal. In RF, we often talk of looking at the amount of power in a signal in two domains (later we will add several more domains): the time domain and the frequency domain. Understanding these domains is useful in understanding exactly what an RF signal is.

If you have used an oscilloscope, then you have seen the time domain. If we look at a signal on the oscilloscope from left to right, what we are looking at is a time span. From bottom to top is the amount of voltage. RF, like your house current, travels in waves in the time domain.

In this domain, we can see that we can measure the position of a wave in degrees as it moves from point to point in time, much the same way we measure an angle moving around a circle. We call this *measurement phase*, and, like a circle, a signal can have a phase of up to 360° (see Figure 1.2). The amount of time it takes for the signal to repeat itself, or get back to its original *phase state,* is called the period, which is a time measurement (time per one cycle). If we inverse this period, we end up with the frequency (cycles per 1 second), in which the unit of measurement is hertz (Hz).

Examining and measuring the phase of a signal is quite important in communications. When we speak of a phase measurement, it is generally with respect to a

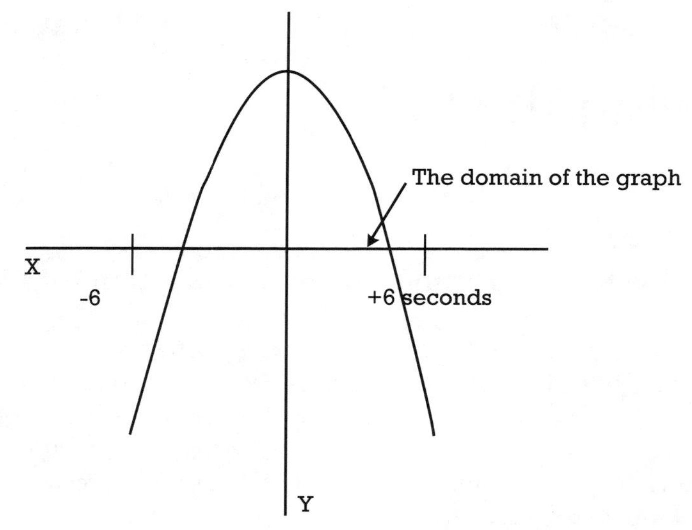

Figure 1.1 Think of the domain of a graph as the X component—in this case, a form of "time" domain.

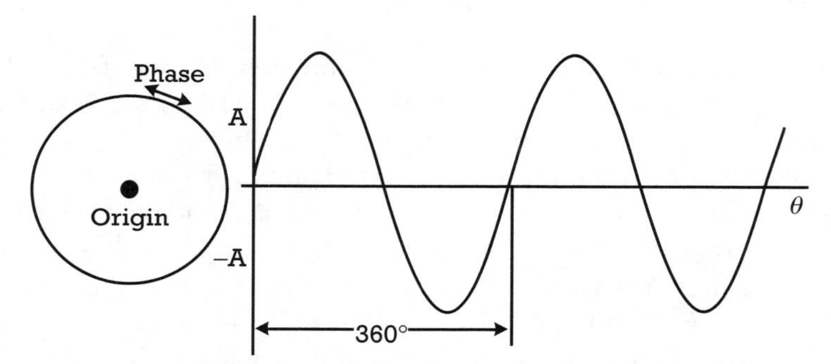

Figure 1.2 A sine wave has 360° of phase, which can be represented by the unit circle (remember, a unit circle is 360° as well).

second signal (see Figure 1.3). For instance, this book will discuss multipath signals later. Multipaths are transmitted signals that take a different path to the receiver. Thus, at the receiver, the two signals have the same information. However, because they take different amounts of time to get to the receiver, they will have different phases. Phase is generally measured in the time domain or via a constellation diagram (discussed later in detail).

In RF design, we are able to separate signals with different frequencies, and hence we can assign different frequencies to different applications so they will not interfere with each other. This is seen much easier in the second domain—the frequency domain.

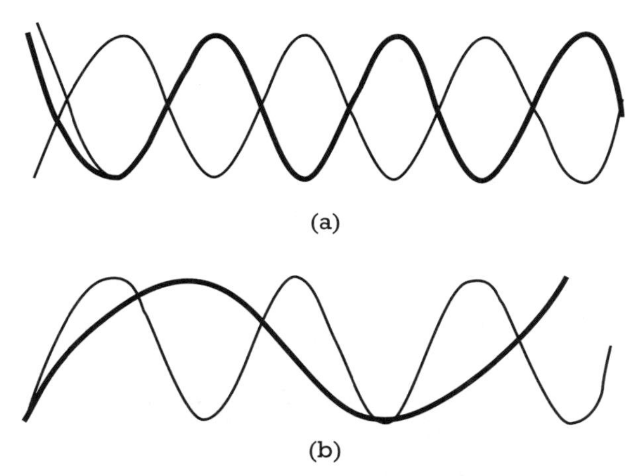

Figure 1.3 Phase and frequency can be seen in the time domain. (a) Two signals in the time domain. Both have the same period and frequency but are 180° out of phase (exact opposite). (b) Two signals: one's period is twice the other—thus, half the frequency.

If you have used a spectrum analyzer, then you have seen the frequency domain. In the frequency domain, we measure the amount of power at each frequency. Chapter 7 discusses spectrum analyzers much more in depth. What is important to understand in Figure 1.4 is that the power of a signal is shown in a frequency domain.

In the frequency domain, it is easy to see how we can divide applications by a specific frequency (see Figure 1.5). For instance, in North America, the cellular phone system is assigned a specific band (around 800–900 MHz) in which to operate, while *frequency modulation* (FM) radio stations are assigned completely different frequencies. With *Advanced Mobile Phone Service* (AMPS), the analog cellular phone system used in North America, each mobile phone conversation is then subdivided and assigned its own frequency band, which is actually 30 kHz wide.

1.3 Power

Once we understand the domain in which we are looking at a signal, of course it makes sense to look at the power component (in other words, the "Y" component of the graph) in the various domains. If you are like most technicians, you probably are very familiar with the oscilloscope. Oscilloscopes usually measure a signal's voltage. In RF and microwave applications, we generally measure a signal's power with one of two units of measure: the watt or the dBm.

While power is always voltage times current (which gives us the wattage), the levels in RF are often so small that it is much easier to represent the wattage logarithmically (see Table 1.1). The dBm is an absolute measurement—0 dBm always

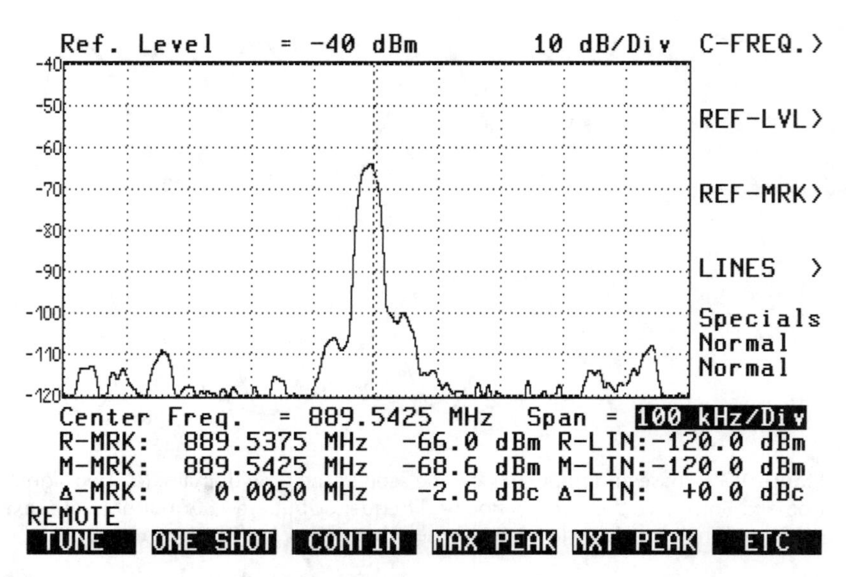

Figure 1.4 A spectrum analyzer shows power in the frequency domain. Note the various setups a spectrum analyzer has in the case displayed to the right and below the signal display. It is important to understand in which domain one is viewing a measurement in order to understand how such setups can affect the measurement.

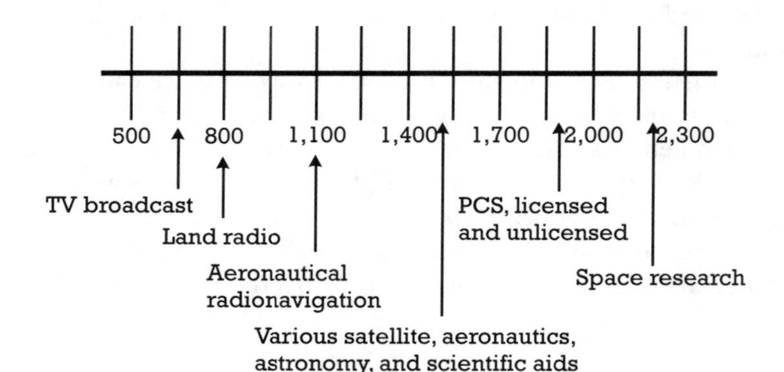

Figure 1.5 The Federal Communications Commission (FCC) assigns frequency bands (in megahertz) based on the application (obviously, this chart is not all-inclusive).

represents 1 mW. When we use dB, it is a relative measurement—in other words, a 3-dB increase would represent a doubling of power, regardless of the starting power level. A very easy way to convert between the two is to remember that 1 mW equals 0 dBm. Then, every 3-dBm increase or decrease is a doubling or halving of power, respectively. You can see why you must have a good understanding of dBM versus watts—the difference in power from 37 dBm to 40 dBm is quite significant, while the difference from 0 dBm to 3 dBm is but 1 mW.

Table 1.1 dBm Versus Watts

Watts	*dBm*
100	+50
10	+40
4	+36
2	+33
1	+30
0.5	+27
0.25	+24
0.1	+20
0.01	+10
0.001	0
0.0001	–10
0.00001	–20
0.000001	–30

We also sometimes represent power levels in terms of their *signal-to-noise* (S/N) ratio. This is important, as many times the ability to receive a signal is not necessarily tied exclusively to the absolute power of the signal at the receive antenna, but the amount of power relative to the noise floor. For instance, if the noise floor in a particular area is particularly high, due to interfering signals or just thermal noise, the S/N ratio will be lower, representing this condition. With the same amount of power in another area, the noise floor might be much lower, meaning a higher S/N ratio and a better condition to try to receive a signal (see Figure 1.6).

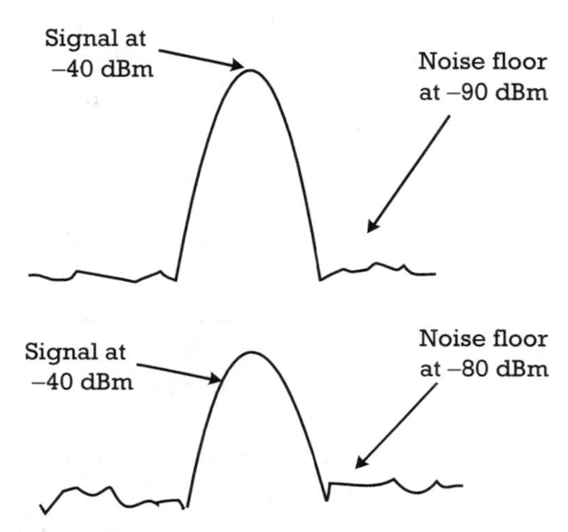

Figure 1.6 Even though the signal level remained the same, the noise floor rose, and thus the S/N ratio worsened.

When it comes to spread-spectrum systems (which will be discussed later), a modification of the S/N ratio is performed to create a different measurement, called E_b/N_0 (pronounced "ebno" by those in the field). E_b/N_0 is often called the digital S/N ratio. It is essentially the S/N ratio with what is called the *processing gain* factored in. This is the benefit derived from using a larger bandwidth to transmit than is actually needed.

1.4 Understanding Propagation

The process of transmitting signals into the air is called signal propagation. Understanding the concepts involved in taking a useful signal and sending it down a transmit path to an antenna and out into the air is an essential aspect of a technician's position. A useful analogy to this process is a water hose. If we turn the faucet on for a moment, we hope that all of the water will get to the nozzle and spray on our vegetable garden. Of course, all of the water usually will not make it. A small leak in the hose might send some of the water where we do not want it. A kink in the line might slow the amount of water down. If it is a large kink, it could stop the water altogether, and in fact send it back to the faucet (backwash), maybe causing more problems. Of course, the best way to make sure everything works well is to use the right diameter hose for the pressure you want.

RF is much the same (except that after a while of efficient watering, we do not end up with plump tomatoes but rather happy customers). We generally have an amplifier, our faucet, on which we attach a cable leading to the antenna, our nozzle. As we send our signal down the cable to the antenna, we hope that all of the power makes it through the route and out the antenna, but there are reasons why this often does not occur: a bad cable that isn't shielded enough, a connector that is not tightened nearly enough, some sort of break in the transmission line, or a flaw in the antenna.

One advantage in the RF world is that by using what is called an RF bridge, we can separate what we are transmitting down the cable and what is coming back the other way (the reflected power—the power that did not make it out shown in Figure 1.7). Chapters 10 to 14 will give more detail on these testing methods.

What also is important to understand is that different frequencies propagate differently. Generally, lower frequencies will travel much greater distances, but they are also more susceptible to physical obstacles, such as a wall. Also, lower frequencies will propagate with a much wider footprint than will higher frequencies; in other words, higher frequencies have more propagation loss than lower frequencies. Higher frequencies are much more directional (hence the use of microwave signals for line-of-sight point-to-point transmission systems). Thus, transmitting 50W at 570 kHz is quite different than transmitting 50W at 800 MHz. In cellular systems, we take advantage of this directionality when we divide cells into sectors. More on this appears in Chapter 2.

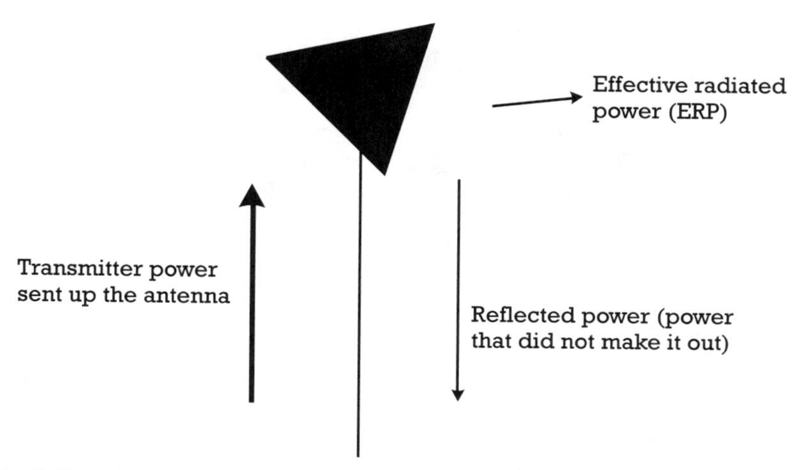

Figure 1.7 Reflected power is a good indicator of antenna and transmit path performance.

The concept of fading also needs to be understood. Fading generally refers to the degradation of the signal at the receiver. There are four main fading effects related to physical conditions:

- Attenuation due to distance;
- Attenuation due to environmental features;
- Raleigh fading;
- Multipath fading.

Naturally, as a receiver moves away from the transmitter or places large objects between itself and the transmitter, the signal level will decrease and lower the call quality. When the subscriber moves a significant distance away from the transmitter, the fading that occurs is called large-scale path loss, as the attenuation is due to the distance. Shadowing generally refers to a type of fading that occurs when the subscriber moves behind a large object or perhaps in a tunnel or the like (see Figure 1.8). These two fades are naturally an obstacle to all wireless systems, although some systems are able to overcome the problem easier than others.

Multipath fading is another type of fading caused by interference between two or more waves leaving the transmitter at the same time, but taking different paths to the receiver and thus arriving at the receiver at different times. This can cause the signals to have different phases as they enter the receiver, and this will cause a destructive process, lowering the actual received level. It should be understood that the multipaths can arrive in phase, and thus they would add constructively—a sought-after condition.

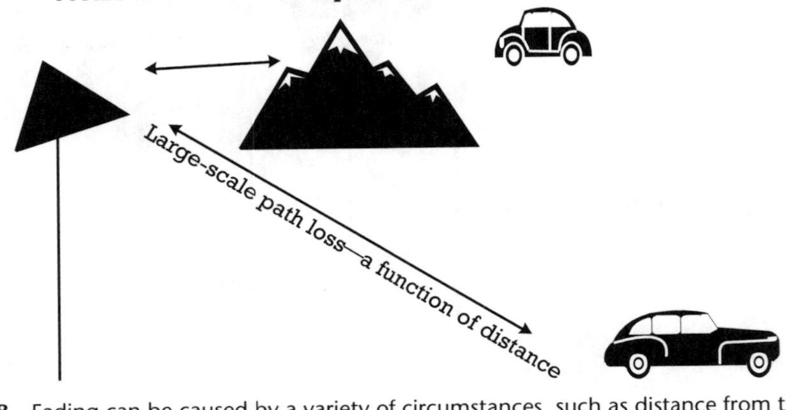
Shadowing, a sudden drop or gain in signal strength, occurs when subscribers pass behind structures
Large-scale path loss—a function of distance

Figure 1.8 Fading can be caused by a variety of circumstances, such as distance from the transmitter or shadowing from large objects like mountains.

If the multipaths could some how be demodulated separately, and then combined coherently, this would substantially improve the system. This is what cdmaOne does with its Rake receiver. Problems in digital systems related to signals taking different paths to the receiver, and thus arriving at different delays, are also often called intersymbol interference (see Figure 1.9).

Fading related to the motion of the terminals (i.e., driving in a car) is often called *Rayleigh fading*. Because of the movement of the handset, the received signal rays undergo what is known as a *Doppler shift*. This causes a shift in the wavelength of the signal, and like the multipath problem described earlier, causes multiple signals at different phases to enter the receiver and degrade (or sometimes improve)

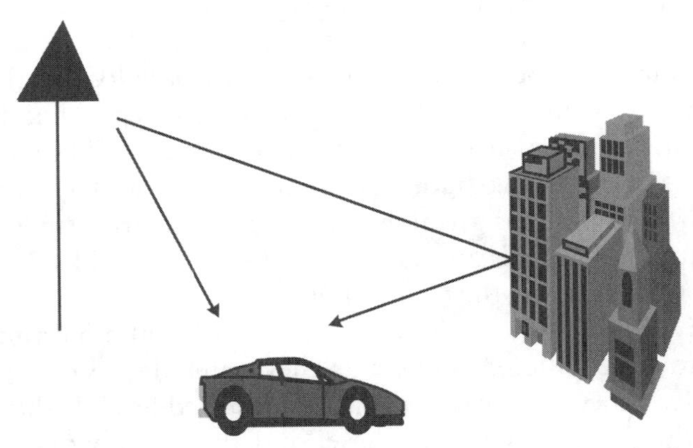

Figure 1.9 Signals can take different paths to the receiver. These multipaths thus take different amounts of time to get to the receiver and can cause problems at the receiver.

reception. The nature of this type of fading causes the changes to occur very quickly, and sometimes this type of fading is called *fast fading*.

Fading can either be frequency selective, acting much like a notch at a specific frequency, perhaps as wide as 300 kHz, or the fades can be flat, causing a problem across a very large bandwidth. One of the best defenses against deep fading is to widen the bandwidth of the signal. Considering that fading can often be frequency selective, one can easily understand how a system that uses a wider frequency band will be able to withstand a fading environment better than a narrowband system (see Figure 1.10). This is one of the main reasons that spread spectrum systems are becoming popular. Also, the fact that fades can be frequency selective explains why the mobiles transmission to the base station may be terrible, but yet the base station connection to the mobile is fine, as these two channels will operate on different frequency bands (separated by either 45 MHz or 80 MHz). These concepts will be discussed in detail later in the book.

1.5 The Transmit and Receive Path

A logical way of looking at a transceiver is to divide the modulation section from the transmit path section. In other words, we can examine the system that modulates and demodulates the information separate from the section that takes the modulated signal and prepares it to be sent out on the air (in the case of the transmitter), and the section that receives the signal off the air and prepares it to be demodulated (in the case of the receiver). The former section will be discussed later in this chapter, as modulation types are discussed. The latter, which we will refer to as the transmit and receive path section, includes connections and cabling from the modulators to

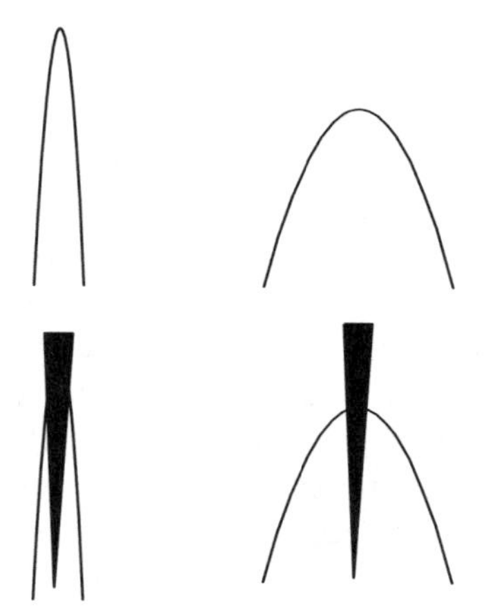

Figure 1.10 A deep fade can occur in a specific frequency band (shown by the black fill). Naturally, the wider the bandwidth of the desired signal, the more resilient it is to these deep fades.

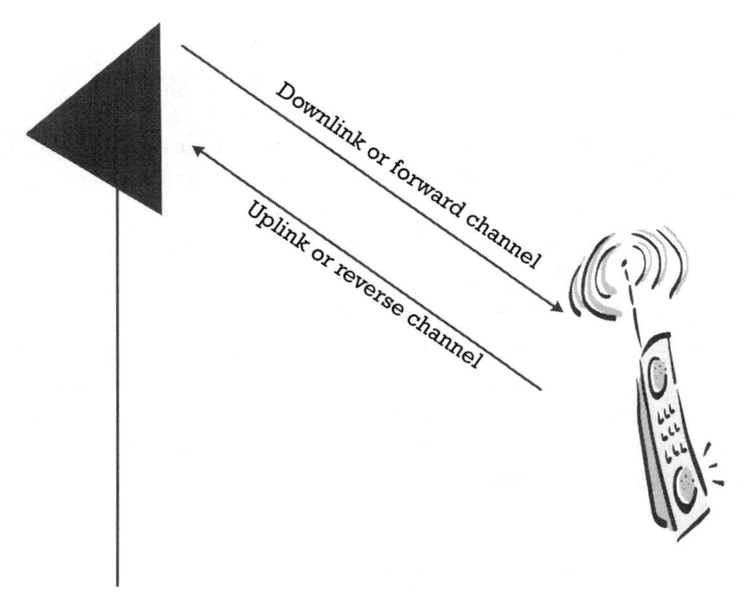

Figure 1.11 In today's high-tier wireless systems, uplink and downlink channels are separated by frequency (frequency-division duplex).

the amplifiers and filters, the connectors and cabling to the antenna, and the antenna itself.

In the main high-tier wireless systems in use today, the mobile transmits on one set of frequencies and the base station transmits on a different set of frequencies. This setup allows for full duplex communications. The mobile transmit band is known as the uplink, or reverse channel. The base station transmit band is known as the downlink, or forward channel (see Figure 1.11).

Filters are self-explanatory. They basically clean up a signal, ensuring that any unwanted power outside of the transmit or receive bandwidth is attenuated. Obviously, filters are frequency dependent, and are generally labeled lowpass, highpass, or selective (and sometimes selectable). These are indications of which frequencies they will allow to pass. As filters are usually passive devices, they do not fail very often in the field, although they can degrade in performance over time. Filters are generally tested by using scalar network analysis (discussed in Chapter 10), which involves sweeping a generator across frequencies at the input and measuring the frequency response at the output (see Figure 1.12).

Amplifiers are used to boost power to a level needed to communicate with other transceivers and to boost received power (these are often called preamplifiers) so that the information can be detected and demodulated. Amplifiers that boost power in terms of watts are known as *power amplifiers* (PAs). Amplifiers come in many shapes and sizes, and they are one of the primary failure areas of transceivers. This is

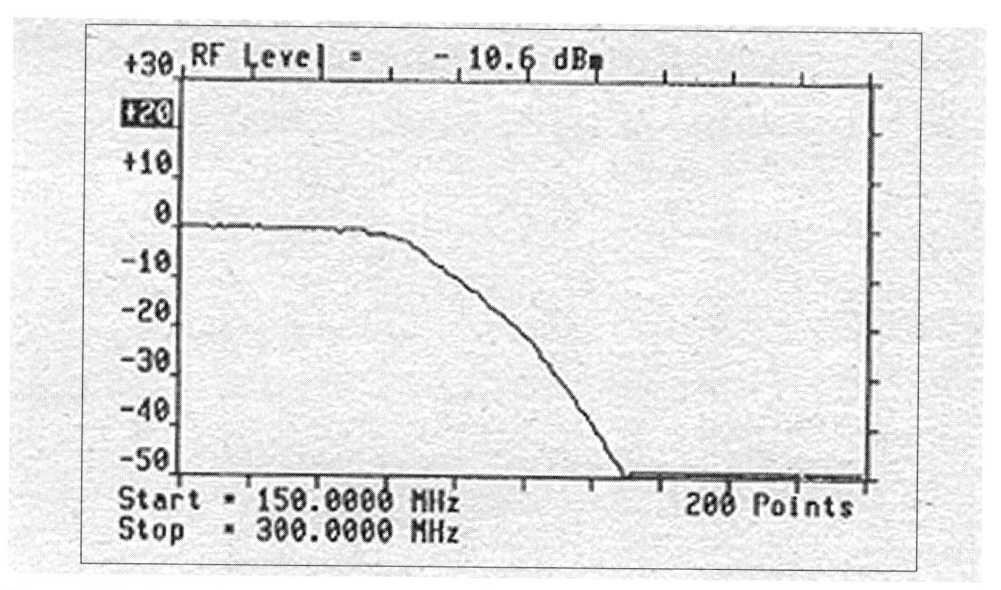

Figure 1.12 Swept frequency analysis is used to measure RF devices in a frequency domain. Here, we see the transmit characteristic of a lowpass filter (note that the higher frequencies are attenuated).

because they are active devices and generally produce substantial amounts of heat in the presence of high-current circuitry.

While amplifier designers try to make their products as linear (frequency independent) as possible, most amplifiers are still nonlinear devices. When RF signals are passed through nonlinear devices, the possibility exists for a form of mixing that causes distortion—more specifically, intermodulation distortion—which can result in poor signal quality or mathematically related interference products at the output of the amplifier. Harmonics (which occur at multiples of the carrier) can also increase or develop in amplifiers, particularly when the signal is overdriven (see Figure 1.13).

1.6 Digital Basics

It may seem out of place to jump from RF propagation to a review of digital technology, but today's complex communication systems are a blend of digital and RF. In fact, the RF portion of these new standards is really the last step in a many-step process of getting the information transmitted. Before we can begin to understand digital modulation, it is important to understand the form of the information we are modulating.

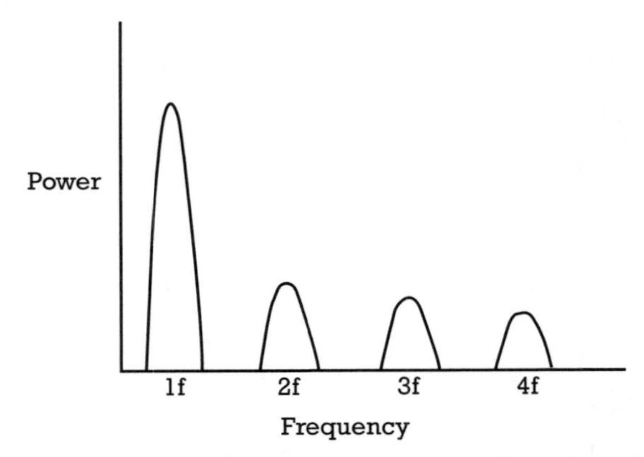

Figure 1.13 Harmonics and intermodulation. Products are usually undesired results of bad devices and are mathematically related to the fundamental frequency.

When we speak of digital, what we are really talking about is the use of ones and zeros to represent data and perform actions. For instance, a computer can use a series of ones and zeros to represent a particular letter of the alphabet, which then allows the computer to store the sequence electronically. It is important to understand that computers speak in digital, and even the "analog" first generation cellular phone system needs the ability to transmit digital information (ones and zeros) in order to relay signaling information.

Usually, this is done by using voltage levels to represent the ones and zeros. Thus, a high voltage might represent a one, and a low voltage might represent a zero. Often, a bipolar digital system is used, in which a high is +1V, and a low is –1V. This allows some easier mathematical manipulation and is generally used in wireless applications.

You can begin to see that ones and zeros could really be represented by any method that has two easily recognizable states—high and low voltage is but one way to perform this task. For instance, we can shift frequency over set periods of time—one frequency for a moment for a one, then another if we want to represent a zero. Another way would be to change the amplitude of the signal depending on the bit to send, perhaps turning the amplitude on for a one and down for a zero. Naturally, some sort of clocking system would need to be in place for this to work.

We could also shift the phase of a signal—one phase represents a one, while another represents a zero. Additionally, we can use relative changes as opposed to absolute positions. For instance, a change of –45° of phase might represent a one and a change of +45° might indicate a zero. This is how we will represent ones and zeros on the RF carrier, which will be discussed later. This form of shifting from one state to another to represent ones and zeros is known as *shift keying*.

Thus, we have *frequency shift keying* (FSK), *phase shift keying* (PSK), *amplitude shift keying* (ASK), and a combination of any two in some systems (see Figure 1.14).

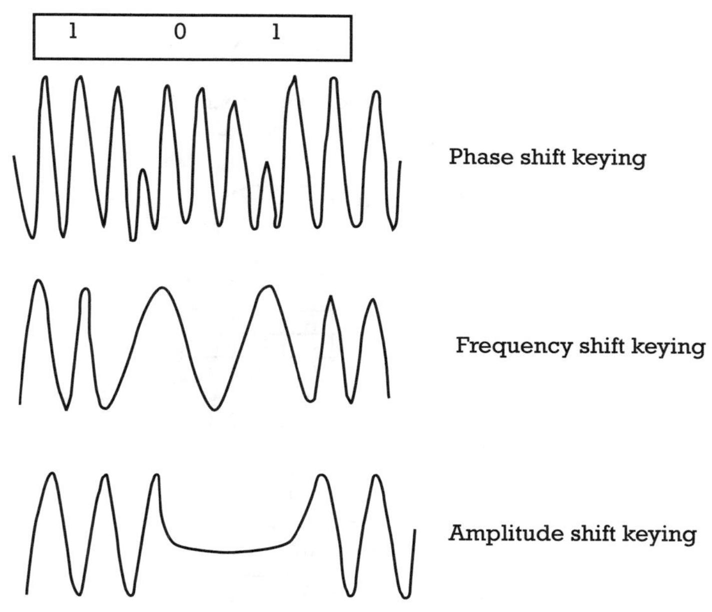

Figure 1.14 Keying certain characteristics to represent ones and zeros.

Up to the point where the data is actually *modulated* on the final RF carrier that will be sent out on the air, data is manipulated in standard *high* or *true* and *low* or *false* voltage digital levels. It is important to review some of the basic digital functions in order to understand how digital sequences are manipulated.

Several logic gates are common in cellular applications (see Figure 1.15):

- *AND gate:* In the AND gate, two trues at the inputs will result in a true output. Any false on an input will result in a false output.
- *OR gate:* In the OR gate, any true value on an input will result in a true out put. The only time it will output a false is if both inputs are false.
- *Exclusive OR (X-OR) gate:* The X-OR gate is very important in digital systems, particularly spread-spectrum systems. The X-OR gate is used to spread sequences (this will be discussed later). A true is output when opposite values are input. In other words, when the input is a false and a true, the output will be true.
- *Inverter gate:* Naturally, the output will be the inverse of the input.
- *Shift register:* A shift register is a logic device used to store a binary value. As it is toggled, it moves its stored value to the output.

Another concept that is important to understand is digital correlation. Correlation is the amount of similarity between two items. In this case, it is an actual

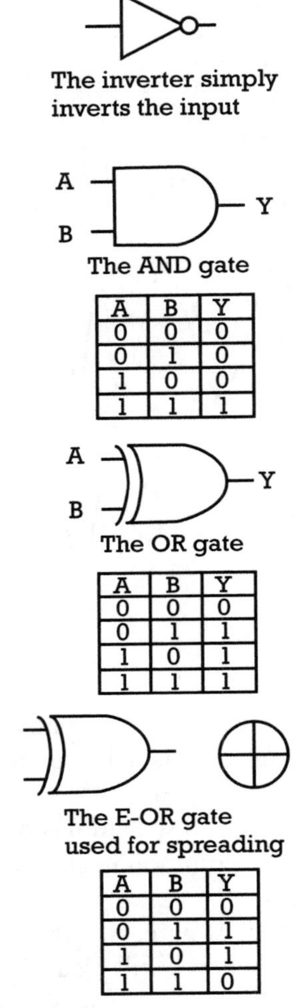

A	B	Y
0	0	0
0	1	0
1	0	0
1	1	1

A	B	Y
0	0	0
0	1	1
1	0	1
1	1	1

A	B	Y
0	0	0
0	1	1
1	0	1
1	1	0

Figure 1.15 Logic gates common in cellular applications.

measurement of the similarity between two digital sequences. If you look at Figure 1.16, you can see how we objectively measure digital correlation. By multiplying the bipolar values, we will end up with a digital waveform product. If we integrate (or find the area of) this resulting waveform, we have a correlation measurement. If 100% of the area is present, it would indicate perfect correlation, or a one. Perfectly uncorrelated signals would be measured as zero. If two digital signals are perfectly uncorrelated, it can also be said that they are orthogonal to each other. You may have heard of Walsh codes in *code-division multiple access* (CDMA). These Walsh codes are actually orthogonal sequences.

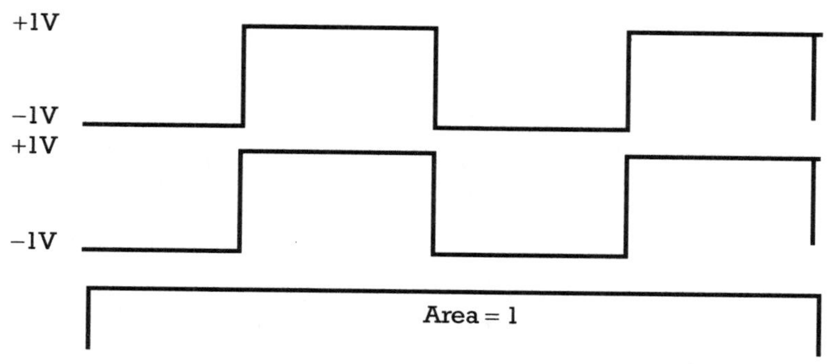

Figure 1.16 The product of two perfectly correlated signals will have an area of one. Any difference between the two signals would result in an area less than one.

1.7 Modulation

Modulation is the process of putting useful information on a carrier that can be transmitted from one point to another. This information can be voice, data, or signaling data. There are several methods of placing information on a carrier, including:

- Amplitude modulation;
- Frequency modulation;
- Pulse modulation;
- Phase modulation.

Two of the more known forms of modulating voice are FM and *amplitude modulation* (AM). Both of these forms of modulation are also known as analog modulation formats, which generally means that analog voice signals are directly modulated on the transmitted RF signal (see Figure 1.17).

It may be useful to review some audio concepts, as it is mostly audio that we are trying to transmit from one point to another. While the human ear can hear a wider frequency range, a toll-quality human speech phone call will generally cover 200 Hz to around 4,000 Hz. In terms of a cellular telephone transmission, this will generally be the audio bandwidth that will need to be transmitted with adequate S/N ratio and low harmonic distortion.

In an AM system, this audio signal will be translated to changes in the amplitude of the transmitted signal, which will be at a much higher frequency. Naturally, when received, these variations in amplitude can then be translated back to frequency variations, thereby reforming the original analog signal that can be sent to a speaker. Because AM systems rely on amplitude changes to transmit the information, they are generally more susceptible to fading and noise than FM and generally require a higher S/N ratio to demodulate. AM signals are also much more affected

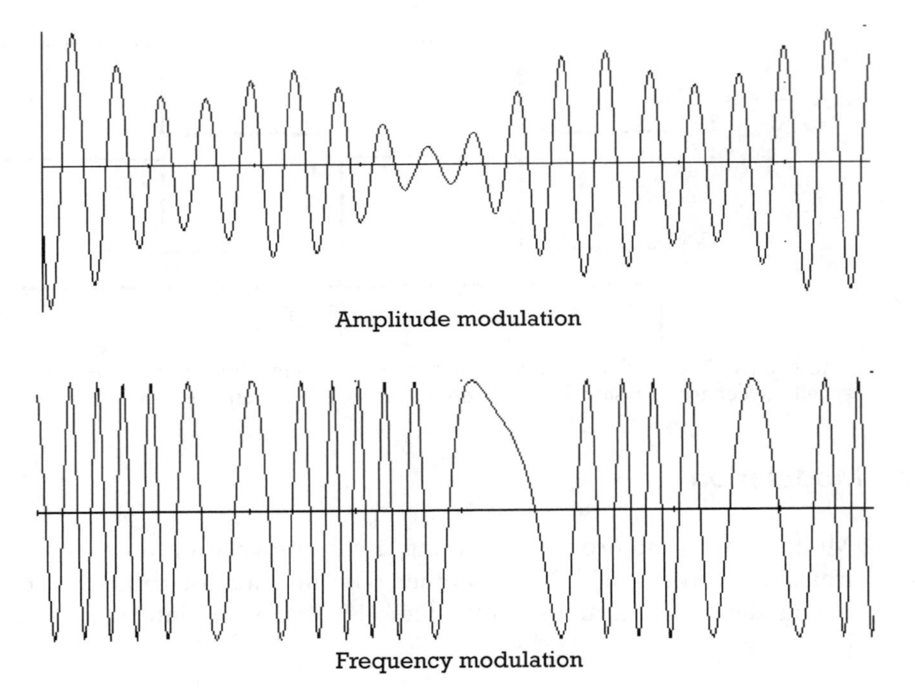

Figure 1.17 FM versus AM (time domain).

by cochannel interference (i.e., another station transmitting on the same frequency). Naturally, FM signals are also affected by cochannel interference, but FM can sometimes select the stronger of the two received signals to demodulate, whereas the AM receiver will have a much more difficult time doing this.

In an FM system, which is the type of modulation used in today's first generation cellular systems, variations in frequency (which relate to the audio to be transmitted) are used to get the information from one point to another.

1.8 Vocoding

In almost all of today's second and third generation wireless communications systems, voice is transformed into digital ones and zeros and modulated on the transmitted carrier using complex digital modulation methods. These methods are substantially different than traditional first generation analog systems.

Because digital modulation implies that the information to be transmitted is already in a digital form (ones and zeros), the first function of the system is to convert the analog voice signals to digital. The basic way of performing this task is to use an *analog-to-digital* (A/D) converter.

An A/D converter samples the voice signal's level and sends a bit stream of ones and zeros to indicate either absolute level or differential steps up or down. The two variables that will affect the quality of the conversion are the sampling rate and the number of bits used to represent steps. There are different standards agreed upon to perform this action.

If you remember, the highest frequency usually found in human speech is around 4 kHz. The Nyquist theorem states that for proper A/D conversion, a sampling rate at least twice the highest frequency in the waveform must be used. Thus, a sampling rate of 8 kHz (which equates to 8,000 samples per second) should be used for human speech. Differences between the created digital waveform and the actual voice waveform are called quantization error.

While this method can make for a high-quality conversation, if the system requires 8 bits for each sample—as is the case for a standard called *pulse code modulation* (PCM)—it would require a data rate of 64 Kbps, which would use too much bandwidth for most wireless systems. Adaptive *differential pulse code modulation* (ADPCM) uses much the same sampling process, but represents the samples differentially (i.e., not as an absolute level, but rather as the difference from the last sample). Using this method, only four bits are needed for each sample; thus, the data rate can get reduced to 32 Kbps. However, this is still too high a data rate for digital systems.

Instead, complex vocoders are used. Vocoders can represent particular human voice sounds with bits, using extremely complex algorithms—essentially using a sort of "codebook." Using this method, data rates below 13 Kbps can be achieved, which are perfect for today's requirements. In simple terms, a system will perform A/D conversion and then run this bit stream through the vocoding process. The vocoding process compresses the amount of bits needed to represent the analog voice via prediction, based on the first few syllables that it "hears" (see Figure 1.18).

Each wireless system generally uses its own vocoder (see Table 1.2). The cdmaOne, for example, uses the *Qualcomm code excited linear prediction* (QCELP) vocoder, while IS-136—the *North American time division multiple access* (NA-TDMA) system, or North American digital cellular—first used *vector sum excitation linear prediction* (VSELP), followed by *algerbraic code excited linear*

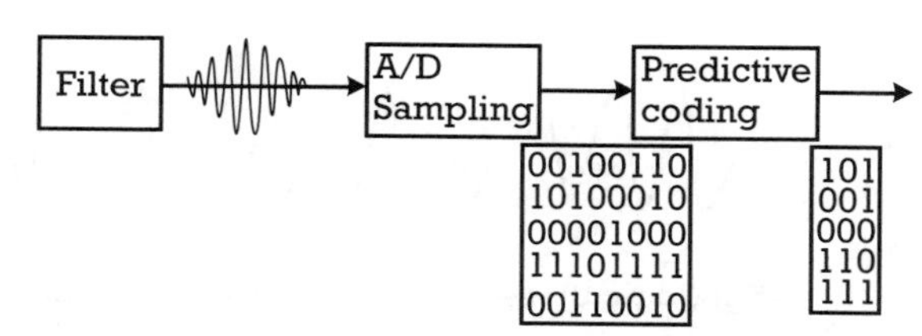

Figure 1.18 Codecs use complex predictive methods to compress the sampled digital data further, allowing less bandwidth to be used.

Table 1.2 Vocoders Used in the High-Tier Systems

	AMPS	*NA-TDMA*	*Global System for Mobile Communications (GSM)*	*CDMA*
Vocoder	Analog	VSELP (original) ACELP	Linear prediction coding with regular pulse excitation (LPC-RPE) (original) Enhanced full rate (EFR)	QCELP
Speech rate (Kbps), including channel coding	N/A	13	22.8 (13 voice)	Variable—8K or 13K at full rate

prediction (ACELP), a higher quality vocoder. Hence, the human voice is turned into ones and zeros and is ready to be modulated on an RF carrier.

1.9 Shift Keying

As mentioned earlier, any method of shifting from one state to another can be used to signify a one or a zero. There are essentially three components of an RF carrier that can be *keyed* this way: the amplitude, the frequency, or the phase—and sometimes a combination of two.

ASK simply varies the amplitude of the carrier between two states, one representing a one, the other representing a zero. Naturally this method is susceptible to the same problems normal AM radios are, and speed is also a problem. Thus, it is not practical for advanced systems (although some systems do use a combination of ASK and another method). FSK, where the frequency is shifted between to states to represent a one or zero, is actually used quite extensively in analog system signaling.

Phase modulation is the system best suited for today's cellular and *personal communication services* (PCS) systems. Naturally, the phase of the carrier is shifted, depending on the data to be sent. The simplest form of this is *binary PSK* (BPSK), where two phase states represent either a one or a zero (see Figure 1.19).

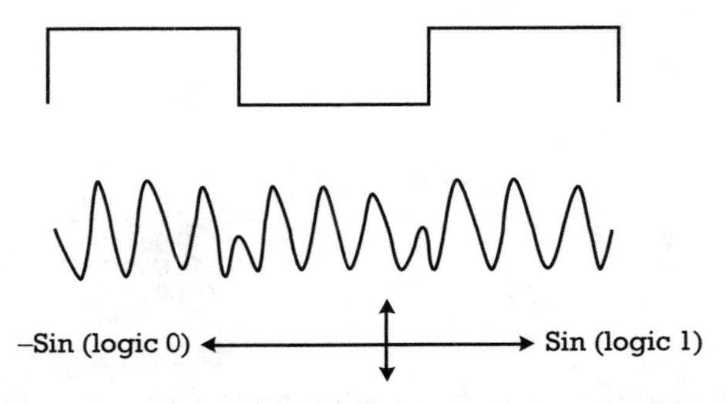

Figure 1.19 Using PSK—in this case, using BPSK, where two phase positions represent either a one or zero.

Phase can be represented in the form of a circular diagram known as a constellation diagram. In this diagram, the phase of the signal is represented by the angle around the circle, while the amplitude of the signal is represented as the magnitude away from the origin (or center) of the circle, with the very center of the circle representing the least amount of amplitude (see Figure 1.20).

All of the main cellular and PCS systems used what is called constant amplitude phase modulation, meaning the amplitude of the signal stays constant during the "decision" points, changing only during phase transitions—and in some cases, not changing at all. Thus, each decision point has two components, phase and amplitude. Deviations from the ideal vector for each decision point is known as the error vector magnitude, which is a combination of two submeasurements—phase error and magnitude error.

In the three widely used cellular and PCS systems in use today, three different types of phase modulation are used. In GSM, *Gaussian minimum shift keying* (GMSK) is used. In IS-136, *π/4 differential quadrature phase shift keying* (π/4 DQPSK) is used. In CDMA, QPSK and *offset quadrature phase shift keying* (O-QPSK) are used for the forward and reverse link, respectively. The 3G formats begin to use higher-order modulation formats, such as *8-phase shift keying* (8PSK), implying eight possible phase states that allow for 3 bits per phase or symbol.

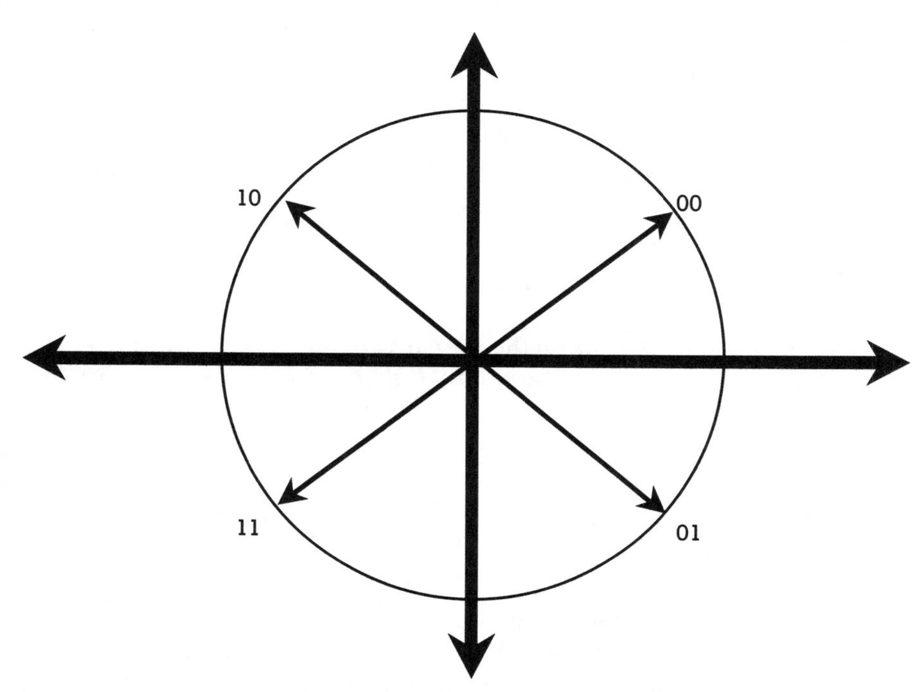

Figure 1.20 A constellation diagram is often used to represent digital modulation. In this case, we are representing quadrature phase shift keying (QPSK), the modulation used in CDMA.

1.10 GSM and GMSK

GMSK is really a type of FSK, which naturally results in phase changes of the carrier. The technique allows the carrier's magnitude to remain constant, with only the phase seeming to shift. The constellation diagram for a GMSK signal, therefore, resembles a circle at constant amplitude around the origin.

The key measurement of this type of modulation, therefore, is phase error, as there is no magnitude error. Many test instruments can plot this phase versus bits, as graphic representations often paint a picture much more clearly. In this way, the technician can quickly see if the modulation is passing or not.

1.11 $\pi/4$ DQPSK and IS-136

IS-136 uses $\pi/4$ DQPSK modulation. There are eight possible phase positions, which are generated by four phase changes: $3\pi/4$, $\pi/4$, $-3\pi/4$, and $-\pi/4$. These four phase changes each represent 2 bits (e.g., 00, 11, 01, and 10), and thus any sequence of ones and zeros can be sent. The demodulator looks for these phase changes, as opposed to absolute phase position.

Thus, the constellation diagram of a $\pi/4$ DQPSK signal has eight decision points, with each decision point having four possible transitions, which will represent one of four 2-bit symbols (00, 11, 01, or 10). Do not be confused by the term decision points. Decisions on the symbols are determined by the transition, and the actual points are simply how the constellation diagram will show problems (if the actual point is wrong, then the trajectory must have been wrong).

1.12 QPSK/O-QPSK and CDMA

The cdmaOne uses QPSK on the forward channel and O-QPSK on the reverse channel. Like $\pi/4$ DQPSK, each symbol represents 2 bits; however, QPSK uses absolute phase position to represent the symbols. Thus, there are four phase decision points, and when transitioning from one state to another, you can see that it is possible to pass through the origin, or center of the circle, indicating minimum magnitude.

O-QPSK is used in the reverse channel to prevent transitions through the origin. Consider the components that make up any particular vector on the constellation diagram as X and Y components (in modulation terms, these would be I and Q). Normally, both of these components would transition simultaneously, causing the vector to move through the origin. In O-QPSK, one component is delayed, so the vector will move down first and then over, thus avoiding moving through the origin and simplifying the radio's design.

1.13 8PSK

Similar to QPSK, 8PSK uses phase locations to determine the pattern. The 8PSK is a higher order modulation scheme, meaning it can transmit more information per shift or symbol. Whereas QPSK transmits 2 bits per symbol, 8PSK transmits 3 bits per symbol. Naturally, this does require a more complex system, as the phase states are closer together.

1.14 Spread Spectrum

CDMA is a spread-spectrum system, but what exactly does that mean? A spread-spectrum system is generally said to be one in which we use more frequency bandwidth than is actually needed to transmit our effective data (see Figure 1.21). Remember that the data rate of the information we are modulating will correlate with the amount of bandwidth used. If we have 13 Kbps of data to transmit, it will generally take 13 kHz of bandwidth. In spread spectrum, we are saying that 13 Kbps will take more than 13 kHz of bandwidth.

By dividing the bandwidth actually used by the information rate, we can calculate processing gain (1.1):

$$G_p = B_w / R \tag{1.1}$$

This processing gain is the ability of the system to fight interference. As you can see, as we increase the bandwidth, the processing gain will increase, and so too will the system's ability to fight interference.

There are several ways to spread a signal. Frequency hopping is one way often used in military applications. It involves changing the carrier frequency being modulated at very fast speeds using complicated algorithms to determine the frequencies.

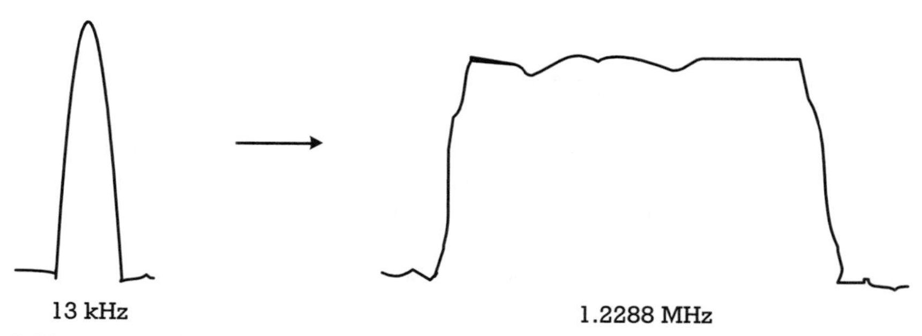

Figure 1.21 In spread spectrum, a data rate that would normally require a small amount of bandwidth is "spread" with a pseudorandom (PN) sequence to a rate that requires a much wider bandwidth.

CDMA uses another method called *direct sequence.* Essentially, we use a PN sequence and an X-OR gate to increase the actually transmitted information.

A PN code or sequence is a sequence of bits that has all the characteristics of randomness, except that we know what the bits are.

As you can see in Figure 1.21, every one bit of actual effective data is X-ORed with the many bits of the PN sequence. The result is that the original 13-Kbps data rate of "effective" data is spread to a transmit data rate of 1.2288 Mbps.

It is important to recognize some of the terms we use to describe ones and zeros. *Bits* generally refer to raw ones and zeros, which represent real information. For instance, the output of an A/D converter would be considered bits. *Symbols* are bits (or groups of bits) that are usually encoded information. For instance, the output of a convolutional encoder, used for *forward error correction* (FEC), which we will discuss in detail later, will be symbols. When we spread data bits or symbols, we call the ones and zeros *chips*. This is why the data rate for CDMA is said to be 1.2288 mega *chips* per second (Mcps) (see Figure 1.22), as the data is actually spread data.

1.15 Conclusion

This chapter should have been an adequate review of or introduction to the basics. The next chapters will become more detailed about the operation of the formats in use primarily in the Americas, often going back over many of the concepts dealt with in this chapter but in greater detail.

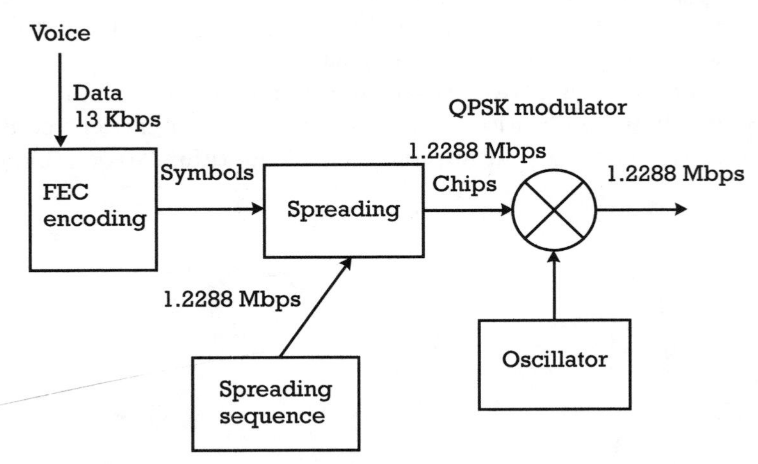

Figure 1.22 Block diagram of a spreading process. Note that the actual information data is only 13 Kbps, while the transmitted rate will be at 1.2288 Mbps.

CHAPTER 2

Cellular Radio Concepts

2.1 The Cellular Concept

With the advancement and shrinkage of today's wireless phones and all this talk of 3G and high-speed data, many people forget a very important concept: these handsets are not really telephones at all, but radios. As radios, cellular phones are subject to the same problems and issues that all RF systems must face. This chapter will review the history of cellular radio communications, discuss the concept of cellular networks and multiple access techniques, and provide the "big picture" of cellular communications.

2.2 The History of Cellular

The concept of cellular communications was developed in response to the limitations of conventional radio services. Although the concept of cellular radio was conceived as early as the 1940s for military applications, the first truly cellular commercial systems were not installed until the late 1970s, with the implementation of the *Nordic Mobile Telephone* (NMT) system in Europe in 1979. The AMPS was implemented in the United States in 1982, and the *Total Access Communications System* (TACS) followed in the United Kingdom in 1983.

As far as mobile phone systems, most would agree the first commercial system was in the implemented in the 1960s with the *Mobile Telephone Service* (MTS), later followed by the Improved Mobile Telephone Service. While these systems were landmarks as far as personal communications went, they seriously lacked the level of performance needed. Even with less than adequate performance by today's standards, as well as a very high price tag, the demand was extremely high. In fact, in 1976, over 3,500 customers were still on a waiting list for the service in New York City. It became obvious that a better system was needed.

The MTS system was a system where one frequency channel was used for the entire system. Thus, the concept was to transmit on the frequency as much power as possible, as more power would increase the cell size. If a user left the coverage area, they had to reinitiate the call on a different frequency channel (see Figure 2.1). If

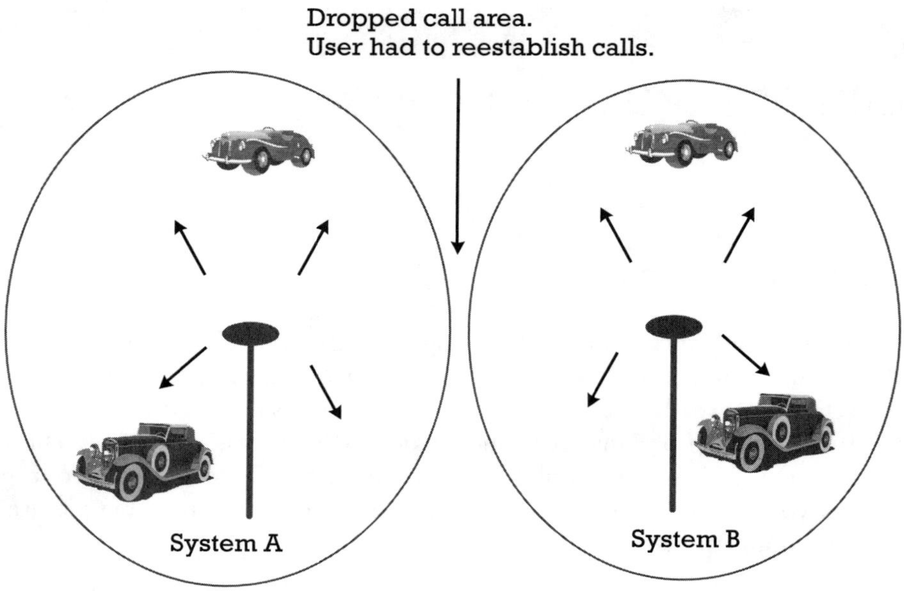

Dropped call area.
User had to reestablish calls.

System A

System B

Figure 2.1 In the beginning, there were no handoffs. The cellular system's size depended on how much power the centralized *base station* could transmit and receive. Users who stepped out of range of one system had to reestablish the call on the next system.

mobile communications was to grow, the first limitation that needed to be overcome was frequency utilization.

The concept of frequency utilization involves two issues. First, it pertains to the availability of actual frequency spectrum free from any interfering coutilization. Second, a method of using the frequencies assigned effectively and efficiently must be developed.

As the twentieth century unfolded, many applications, from television and radio to air-to-air and satellite communications, began eating up this new domain of frequency spectrum. In the United States, the FCC needed to isolate and assign a band for this new and growing communications trend. As frequency goes up, it becomes more difficult to develop cellular communication systems. Many would agree that 10 GHz would be the upper limit. Thus, there is indeed a finite amount of bandwidth available, yet essentially an infinite amount of applications our best scientists can come up with.

The original MTS systems operated at 40 MHz, 150 MHz, and 450 MHz. The FCC had already assigned most of the frequency bands below 400 MHz to other applications. As one of the pioneers in communications, the Bell System had proposed using the 800-MHz band as early as 1958.

Up until the mid-1970s, the 800-MHz band was being used by educational television. However, the advent of cable television lessened the load on the spectrum, as many of these programs could move off of the airwaves and onto the cable lines. In 1974, the FCC allocated 40 MHz of spectrum in the 800-MHz band, originally allowing one cellular operator in each market.

In 1980, the FCC changed its mind on the one-carrier-per-market concept and decided on allowing two carriers to increase competitive pressures. The frequencies were then assigned in 20-MHz blocks designated A and B band. Because the FCC realized that local wireline providers would be a primary player in cellular communications, the B band in each market was assigned to wireline companies, while the A band was reserved for nonwireline companies. Thus, utilizing the AMPS protocol, there were 333 30-kHz frequency channels per carrier per market, numbered 1 to 666.

In 1986, in order to increase capacity for the ever-growing cellular industry, the FCC decided to add bandwidth to each band. This accounts for the disjointed frequency channel numbering seen in the frequency channel chart for cellular (see Figure 2.2). The enlarged spectrum system is sometimes referred to as *extended Advanced Mobile Phone Service* (E-AMPS).

While we will discuss specific properties of the air interfaces later, it is important to first understand the channel numbering structure. In the cellular band, each channel is 30 kHz wide (as you will see later, each AMPS channel requires 30 kHz).

In the 1990s, cellular exploded in growth, and we wondered how we ever lived without it. As the demand for service increased, carriers began to have problems delivering with their finite spectrum and technological resources. Digital formats, which could increase capacity, began to emerge, and the FCC set out to allocate further spectrum. Two digital standards arose, NA-TDMA and cdmaOne. Both

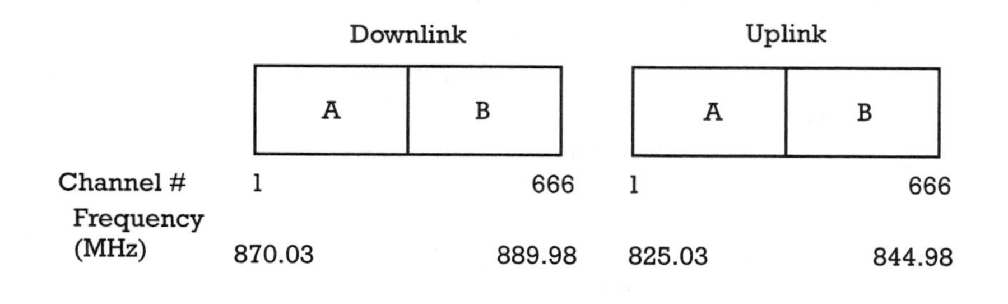

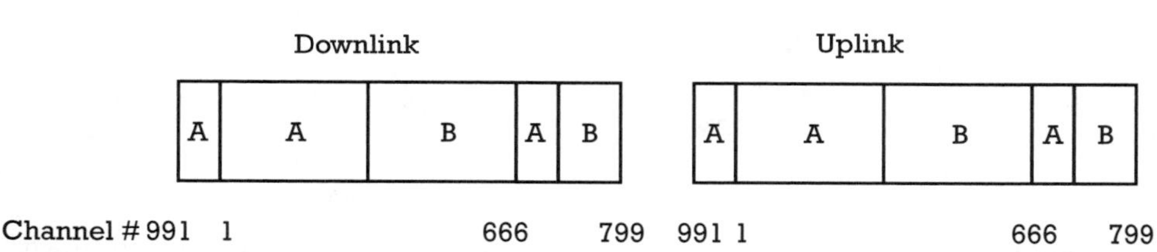

Figure 2.2 The cellular channel structure, a bit convoluted after the addition of spectrum.

systems were dual-mode standards, allowing a seamless transition from analog to digital.

As the demand for more services and more competition grew, what emerged was a new frequency band allocated for PCS. The licenses for these bands (designated A–F) do not specify which technology could be implemented, and they were auctioned off (at what many felt were outrageous prices). The use of these PCS bands has become widespread, and the competitive forces associated with two to four new mobile phone carriers in each market have caused a significant drop in the cost of wireless services, with a significant increase in the performance and benefits.

The PCS band uses 50-kHz channels as far as its channel numbering system. The channels of several of the formats using this band have occupied bandwidths much larger than 50 kHz (i.e., CDMA uses 1.2288 MHz and GSM uses 200 kHz), so channel numbers refer to the center frequency of the carrier.

In Europe, most nations also adapted some form of an analog system early on. One of the first systems was the NMT system, which covered Denmark, Finland, Norway, and Sweden. If a user wanted to roam, often even within this system but to a different nation, a different handset was needed. In the 1980s, Europe faced the same phenomenal growth seen in the United States. As its capacity problems began to increase, and as a more open and mobile European society emerged, the demand for a next generation increased.

The European Community decided the best option was essentially to start from scratch. They allocated a 900-MHz band across the continent and approved a new digital standard, GSM, which specified both the air interface as well as the signaling specifications. Unlike the American next generation goals, which focused primarily on seamless integration into the already extensively deployed AMPS system, GSM was designed with standardization and roaming as the key features, with no need to be backward compatible. Thus, while Europe converged on a single standard, the United States split from its original AMPS to *time-division multiple access* (TDMA) and CDMA, as well as keeping, and in some cases even expanding and improving, the existing analog AMPS networks (see Figure 2.3).

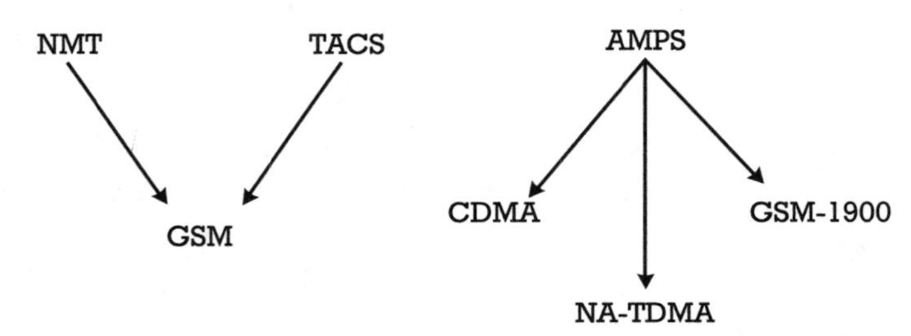

Figure 2.3 While Europe converged on the single standard of GSM, North America diverged from AMPS into NA-TDMA, CDMA, and GSM-1900.

In the 1990s, Europe also established its own new band for what it called personal communications networks. The band, now called DCS 1800, utilizes an upbanded GSM network in the 1,800-MHz band.

Japan also established its own system along the same time frames. Japan developed *Pacific Digital Cellular* (PDC), which is very similar to NA-TDMA. A lower tier format called *personal handyphone system* (PHS) is also very widespread in Japan. PHS is primarily designed for pedestrian, low-speed use and has been enormously successful, to a great extent, as a substitute for landline service—this would be called a *wireless local loop* (WLL) system. Japan has also deployed a successful CDMA network. In addition, Japan allocated a separate frequency band specifically for *wideband code division multiple access* (WCDMA). Thus Japan was the first nation to deploy a commercial WCDMA network, albeit with some initial deployment issues.

Table 2.1 lists just a few of the analog formats used in different regions. All were similar, and none of them allowed roaming from system to system.

With the 2000s, many thought cellular would evolve quickly again, as customers would look for wireless Internet at *digital subscriber line* (DSL) or cable modem data rates. In addition, wireless operators were looking for more voice capacity out of their spectrum. Hence, many saw a third generation of wireless formats as the next big thing in cellular communications. With a global economic slowdown, however, and financial struggles of many operators, many of whom were already struggling from the cost of their original spectrum licenses, the 3G deployments have been significantly slower than expected.

Operators, however, still must plan their roadmap for the future. As you will see in later chapters, CDMA was designed right from the beginning with an evolutionary path to 3G services. Engineers designed the *general packet radio service* (GPRS) and *enhanced data for GSM evolution* (EDGE) formats to allow stepping stones to the true 3G format: WCDMA. NA-TDMA, which saw widespread deployment across the Americas, did not seem to have a clear migration path to 3G or high-speed data rates. Although there were several viable formats in early stages that would allow for such a transition for NA-TDMA, economies of scale did not make most of them realistic for deployment. As a result, most of the major NA-TDMA

Table 2.1 A Few of the Analog Formats Used

Country/Region	*Format*
United Kingdom	TACS
Canada	Aurora
Scandinavia	NMT
Netherlands	ATF2
United States	AMPS
Austria	Autotelefonnetz C
Hong Kong	AMPS, TACS, Japanese NEC

operators have decided to move towards GSM and hence take on GSM's evolution path.

As mentioned earlier, there are two issues associated with frequency utilization. The first, spectrum allocation, was taken care of in the previous paragraphs, where frequency bands are assigned by the governments. The second issue, and certainly the more difficult issue to address, is creating a system that efficiently uses the spectrum that is assigned.

2.3 The "Cell" in Cellular

In transitioning from the conventional MTS-type wireless systems to the first generation cellular systems, engineers needed to develop ways of using their frequency channels effectively. If only one frequency could be used at a time in a specific geographic system, then only 333 calls would be able to be made at any given moment per carrier. In addition, the mobile user would not be able to stray out of the covered cell without dropping the call, unless a complex repeater system was installed; this was hardly adequate for markets the size of New York and Chicago. Thus, the concepts of frequency reuse, cells, and handoffs were born.

Naturally, the geographic boundaries of a mobile communication system at a specific frequency will be determined by the amount of power that can transmitted and received by the base station (the fixed infrastructure) and the mobile terminal (henceforth referred to as the handset). Obviously, external factors can limit the coverage as well, such as terrain, interference, and frequency, which naturally effects the propagation characteristics (see Figure 2.4).

In cellular radio (from this point forward, cellular radio refers to all systems that utilize a cellular geographic layout, be it at PCS or cellular frequencies, regardless of the format), the marketing concept of the hexagon was established to help visualize particular geographic areas. Each of these hexagons will represent a base station sector. While the hexagons make planning and visualization easier, frequency coverage will not resemble this pattern as all—in fact, the ideal shape would be a circle, while the typical shape is generally more like a patternless blob (see Figure 2.5).

The advent of microprocessor technology, as well as improved miniaturization of RF technology, was a key factor in allowing for a system with better frequency utilization. The ability of a mobile user to *hand off* without dropping the current conversation and reestablishing the call meant that adjoining cells could use different frequencies. As long as the cells were far enough apart so as to not to interfere with each other, frequencies could be reused.

Planning these frequency channel plans thus became a large part of the cellular industry. The design engineer must estimate decide the distance required before reusing frequencies. Too much distance will naturally decrease spectrum efficiency. If base stations are too far apart, they will cover too large an area and thus there may

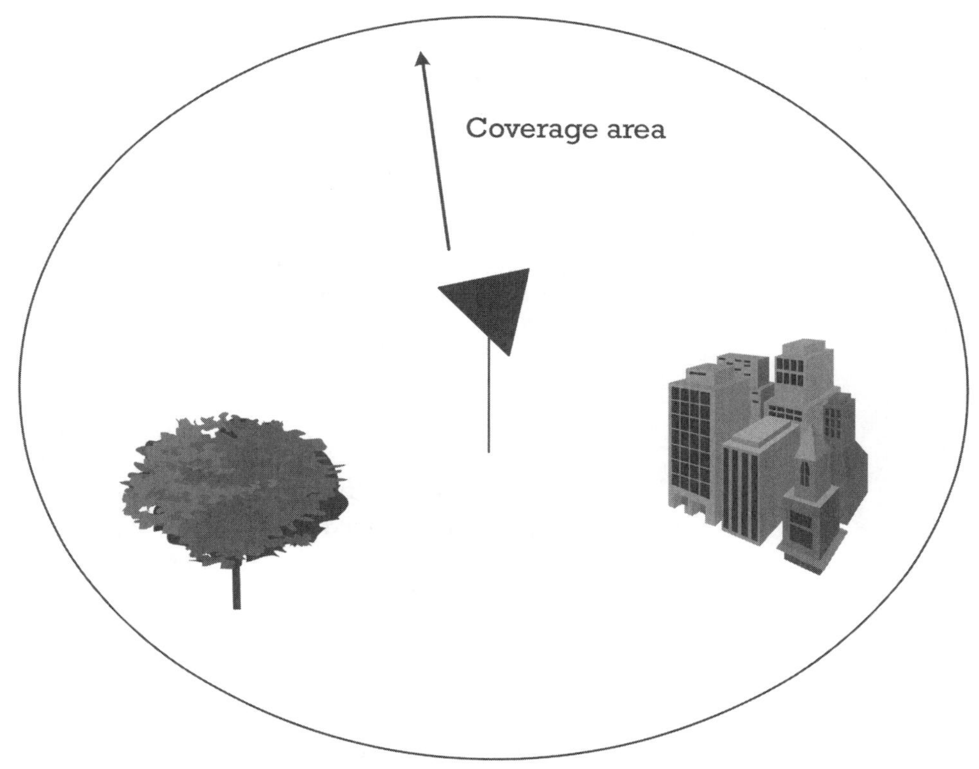

Figure 2.4 The coverage area of a cell depends on several factors, including the power output of the transmitter, the performance of the receiver, the terrain (including some dynamic, seasonal terrain aspects, such as foliage), as well as the amount of interference in the system.

be too many users in a large geographic area to be serviced by only one base station. Too little distance will cause cochannel interference—that is, when the signal from two base stations using the same frequency reach the mobile's receiver (see Figure 2.6). In systems that use different frequency channels for each user, eventually these frequency channels must be reused, with the idea that they will be reused at a sufficient enough distance from each other that they will not reach each other.

While channel planning is generally is generally a high-level engineering task, often times base station technicians must interact as base station hardware is blamed for what is actually a network design issue. Also, handset technicians are often faced with upset customers who seek to blame the handset (and very often the handset technician) for coverage issues. The ability of technicians to have a firm grasp of network layout and coverage issues is a tremendous advantage for the technicians as well as the carrier. The latter portion of this chapter will discuss many of these issues, most of which are not necessarily issues assigned to technicians but are very important for technicians to understand.

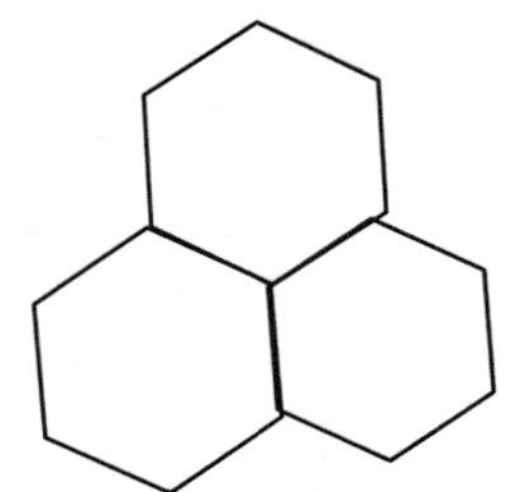

No book on cellular would
be complete without hexagons.

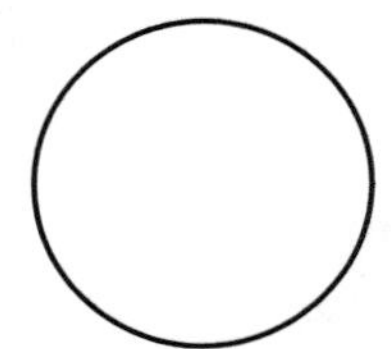

Alas, the real-world cell.

Figure 2.5 The marketing hexagons, the ideal circle, and a more realistic cell.

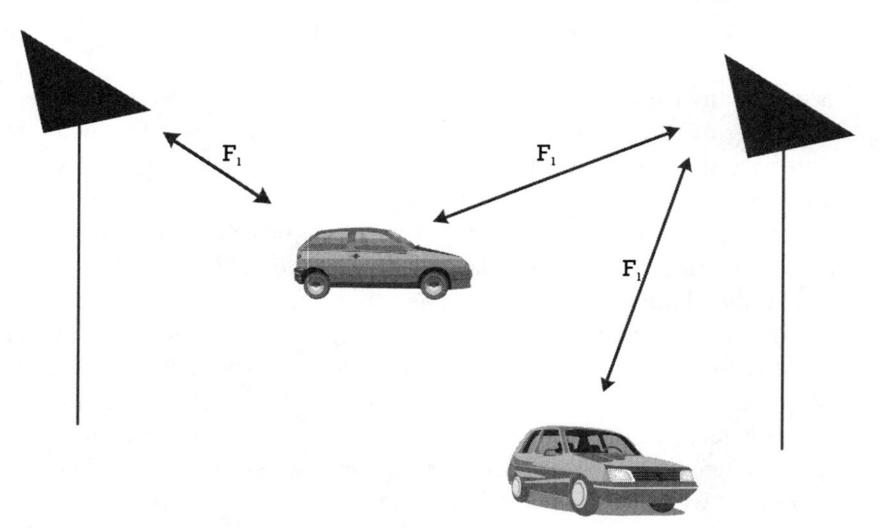

Figure 2.6 Poor frequency plans (i.e., using the same frequency channel too close together) leads to a problem called cochannel interference.

2.4 The Typical Modern Network

After Chapter 1, you should be familiar with the terms base station and mobile station. This section will review the components of a cellular network in order to familiarize you with network terminology.

A cell is a base or land station. In most urban and suburban locations, these base stations are typically sectorized. This means that three separate antenna systems transmit and receive to cover the cell.

The AMPS system clearly defined this base station's air interface (i.e., the interface between the base stations and the mobile stations via RF). Many of the terms are shown in Figure 2.7. Thus, any AMPS mobile phone is compatible with any AMPS base station, even if the base stations are made by different manufacturers and operated by different carriers. The problem with the AMPS standard is that it did not specify the operations behind the air interface. This made certain operations, specifically roaming ability, quite difficult.

In Europe, the ability to roam was a top concern when they were developing their next generation system, GSM, the layout of which is shown in Figure 2.8. Hence, the procedures for setting up roaming systems, as well as the technology for the interfaces and systems, were standardized. North America, sensing the need for standardization, borrowed much of the terminology and protocol details from the GSM system and created the IS-41 standard. This explains one of the reasons that there seems to be several names for the same thing in North America [AMPS used the term *land station*, while GSM used *base transceiver station* (BTS), and others used *base station*].

IS-41 specifies two databases used to assist the network in setting up roaming calls. The first is the *visitor location register* (VLR). The VLR stores the important

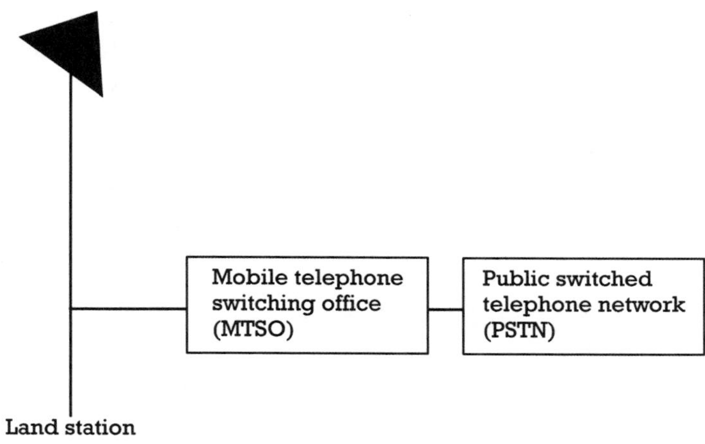

Figure 2.7 Terminology and architecture of a typical AMPS network.

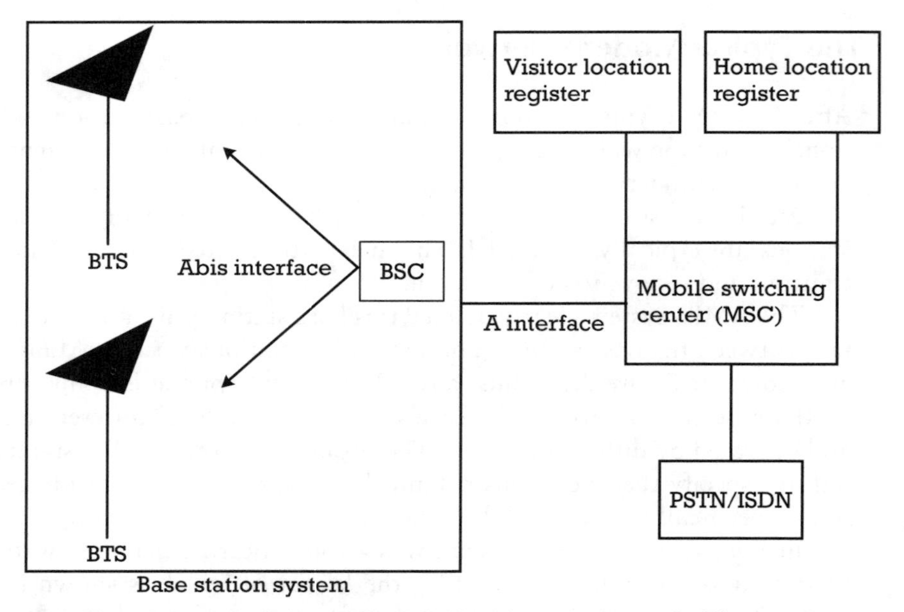

Figure 2.8 The GSM standard network layout. (IS-41 borrowed several of the terms.)

information about each user that it knows is in its coverage area. In this way, the *mobile switching center* (MSC)—formerly the *mobile telephone switching office* (MTSO)—knows who and where the terminals in its area are. The second database is the *home location register* (HLR). The HLR contains the subscription information for subscribers. Using these two databases, the MSC can quickly set up calls, whether incoming or outgoing.

The MSC then links up with the *public switched telephone network* (PSTN), which, of course, is the outside world. Communication between the BTSs or base stations and the MSC is usually in the form of a T1/E1 line or a microwave link. Communications from the MSC out to its databases generally relies on common telecommunications networks, very often the *Signaling System 7* (SS7), which is a layered network protocol.

Specific details on network architecture, which will be different for each technology, will be included in the later technology-specific chapters.

2.5 The Concept of Multiple Access

If a cellular system only allowed one call per system at a time, it really wouldn't be very practical. Thus, cellular designers need to implement methods of multiple access (i.e., allowing multiple conversations simultaneously). Also, because cellular and PCS phones are trying to be as similar as possible to regular landline phones, they must be full duplex (i.e., both parties must be able to talk simultaneously), unlike a traditional "walkie-talkie," where one person would trigger their microphone to talk at a time (half-duplex).

The concept of duplex must first be understood. This is the way we separate the forward and reverse path, as they obviously cannot share the same time or frequency. Almost all high-tier wireless standards use some form of *frequency-division duplex* (FDD). FDD simply means that two separate frequencies are used for the forward channel (from the base station to the mobile) and the reverse channel (from the mobile to the base station).

You do notice, however, that you can hear in the speaker what you say into the microphone during a conversation. This is due to a loop-back of the voice after it has been transmitted to the base station. Most of the high-tier wireless formats use FDD, while several cordless formats, such as *Digital European Cordless Telecommunications* (DECT) and PHS, use *time-division duplex* (TDD), in which the same frequency is used for the forward and reverse channel, but different time slots are assigned. In any case, modern wireless systems are typically full duplex; thus, the need for saying "over" at the end of each transmission, so the other end of the conversation knows it is his or her turn to talk, is indeed gone.

That said, with spectrum as tight and expensive as it is, TDD becomes an attractive alternative, as only one frequency band, which is shared by the uplink and downlink, is needed. There are some limitations to TDD, but the TDD versions of the 3G WCDMA standard will be deployed.

Dividing each user, or channel, by frequency is a logical and simple solution (see Figure 2.9). Thus, most multiple access systems use some form of frequency-division multiple access (FDMA). Thus AMPS, the analog system used in the Americas, is an FDMA/FDD wireless format.

Today's digital systems use a combination of FDMA with another multiple access technique, TDMA (see Figure 2.10). TDMA involves separating users into different timeslots and then adding at different frequencies these "time-divided"

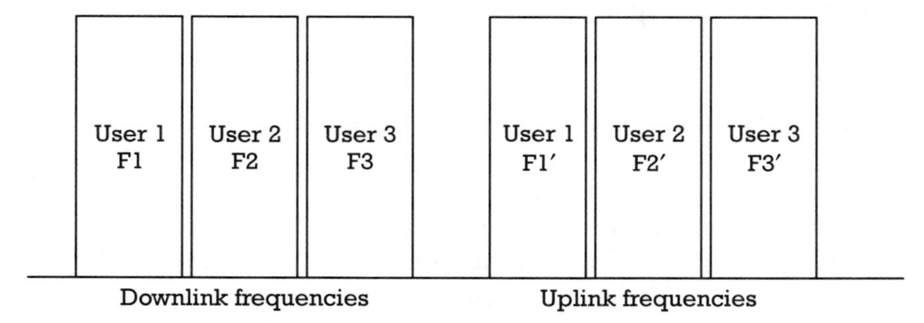

Figure 2.9 FDMA involves separating users by frequencies—each user gets his or her own frequency set: one frequency for the uplink, and a corresponding frequency for the downlink.

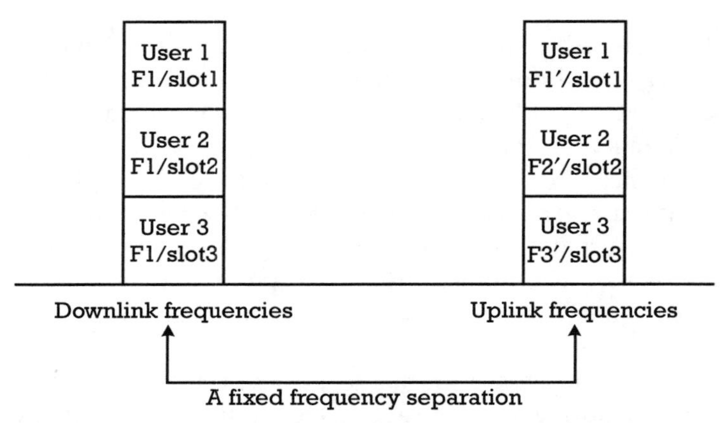

Figure 2.10 TDMA/FDMA involves separating users by time slots, as well as frequencies. Each user gets assigned a frequency and then a time slot when they are allowed to transmit.

channels. GSM places eight time channels on each 200-kHz frequency channel, while IS-136 (NA-TDMA) places three (full-rate) channels on each 30-kHz frequency channel. Half-rate systems are also being deployed for both GSM and NA-TDMA, which means the number of time slots, and thus conversations, per frequency channel is doubled. Nomenclature used to describe both would thus be TDMA/FDMA/FDD.

Naturally, a TDMA system must compress the voice information so that it can fit in a smaller slot of time, later stretching it to normal after it has been transmitted and received. While this oversimplifies the process, the concept is what is truly important.

CDMA involves separating users by a specific Walsh code, while all of the users transmit at the same time and on the same frequency band. There are 64 channels in the same 1.2288-MHz channel (although at least three code channels must be used for overhead). If additional capacity is needed in an area, another frequency band can be allocated, thus making it "sort of" FDMA as well (see Figure 2.11). IS-95 and J-Std 8 are CDMA/FDD systems.

2.6 The Key to Truly Mobile Communications—The Handoff

As mentioned earlier, setting up geographic cells and systematically reusing frequency channels necessitates a method of transferring the call from one base station to the next without dropping the call and requiring reestablishment. The solution is the handoff. While the specific technology chapters will discuss how handoffs are achieved in each format, this section will provide a brief overview of the concept.

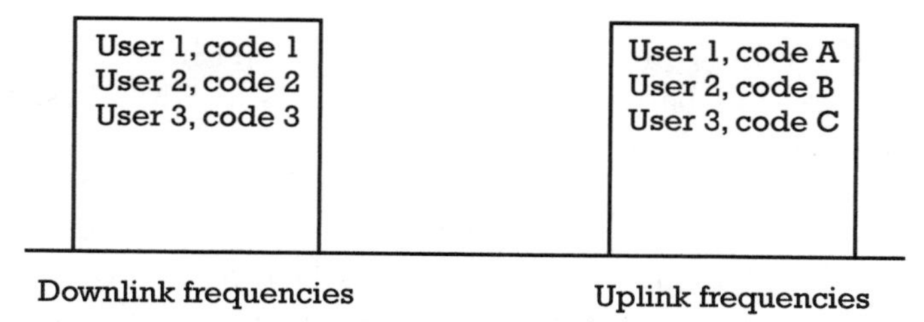

Figure 2.11 CDMA does not assign time slots, but rather assigns codes. Users transmit on the same frequency at the same time, separated by unique codes. Additional frequencies can be assigned to increase capacity, so it can be FDMA as well.

Naturally, if a handset is to switch from one base station to another, the key to the process would be in determining the right moment to switch (see Figure 2.12). Analog, GSM, and NA-TDMA use hard handoffs. This means that the handset will first disconnect from the current base station and then change frequencies to the next base station—in other words, it is a break-before-make handoff. CDMA adds the soft and softer handoff functionality. While these handoffs will be discussed in detail later in the book, they basically refer to a make-before-break action. In other

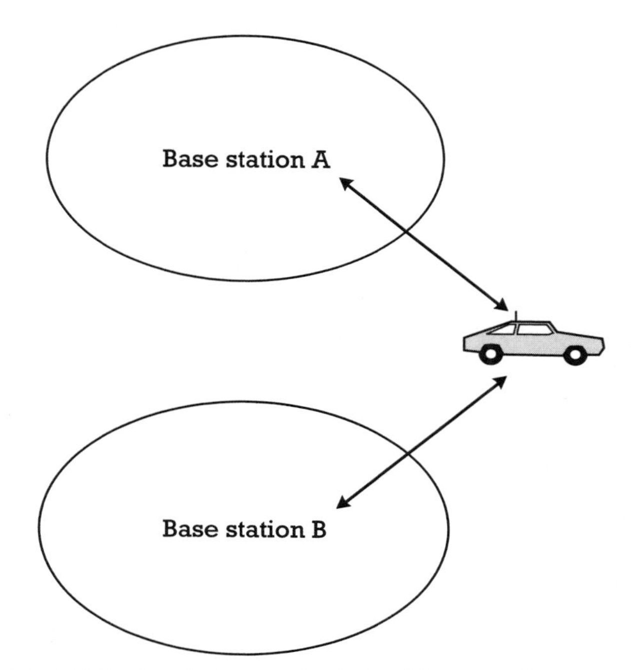

Figure 2.12 In a handoff, the handset reports back signal strength of the neighboring base station to help the system know which base station to use. In CDMA, the mobile can actually use more than one base station at a time (known as a *soft handoff*).

words, the handset will actually listen to both base stations simultaneously and then drop the degrading base station. The fact that all CDMA base stations use the same frequency is what allows this process.

In the analog systems, the process of the handoff is controlled by the base station controlling system. Base stations typically use location receivers, which can measure the received reverse channel power. When the received power drops below a certain threshold, either an absolute power level or the *carrier-to-interference ratio* (C/I), the BSC will initiate and control a handoff.

Naturally, this process requires intensive and complex calculations. The handoff must occur at the right moment, or many calls will drop. Handoffs that occur too soon or too late will mean the handset will not have enough power to keep the call up.

Because handoffs are one of the primary causes of dropped calls, and thus customer complaints, the designers of the second generation digital systems needed to improve handoff performance. As TDMA systems divide their transceiver time into time slots, they have free time to monitor other channels.

Thus, the TDMA handset can assist in the handoff process. This *mobile assisted handoff* (MAHO) allows the system to make better determinations on when the handoff should occur and which base station would make the best candidate to whom to hand off. By continuing to monitor the *receiver signal strength* (RSS) measurements from the base station, as well as monitoring the *received signal strength indicator* (RSSI) from the mobiles, the BSC can have complete information on the forward and reverse paths in order to make determinations and thus better decisions on when and where to complete a handoff. Each individual technology chapter will discuss the handoff process in detail.

2.7 Sectorization

As the demand for mobile service increases, the need to increase system capacity and performance grows as well. Because we are dealing with a finite amount of spectrum for channels, and creating and implementing new technology is expensive and time-consuming, the concept of cell splitting evolved. This process is only possible because of virtually seamless handoffs.

By dividing the cell geographically, usually into 120° sectors, a carrier can increase its coverage and improve the overall quality of the system, as directional antenna systems are usually better performers. In WLL systems (i.e., wireless systems designed to substitute for landline phones), mobility (and thus handoffs) is not as important, and often six sectored base stations are used.

Generally, base stations use the same hardware, assigning individual radios to each sector as needed. This substantially reduces the amount of hardware needed and allows the carrier to aim and design sectors to solve coverage issues (see Figure 2.13). Thus, sectorization in itself does not improve capacity (except in the case of

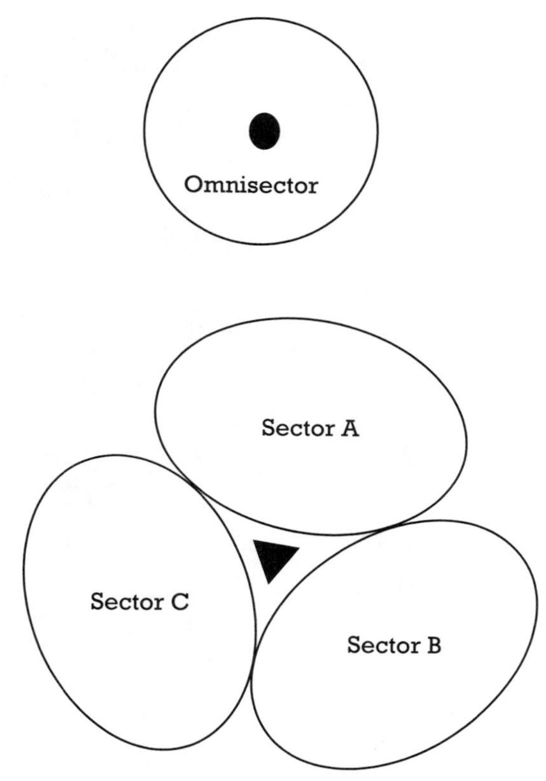

Figure 2.13 Sectorizing a base station can help improve coverage and better utilize equipment.

CDMA), but the use of directional antennas and the ability to better utilize frequency channels make it very beneficial to the network.

In areas where capacity and network layout is not an issue (e.g., in very rural areas), omnidirectional cell sites are still used extensively. In addition, the new breed of microcell or picocell sites, designed to add coverage to malls and office buildings, also often are omnidirectional.

2.8 Antennas

Configurations of cell sites can vary depending on the manufacturer of the infrastructure. Each base station has location and setup radios in addition to the voice radios. Very different patterns can be used, but a 120° sector, using directional antennas, usually has one transmit and two receive paths (see Figure 2.14). These receiver antennas are usually called *diversity antennas*. In the past, this usually meant there would be three antennas per sector. In addition, as more channels are

Figure 2.14 A typical cellular/PCS antenna system; actually, two colocated systems are shown. Note the three antenna elements per sector, generally two receive antennas and one transmit antenna.

added, oftenanother transmit path is needed, meaning there would be four antennas per sector. With localities starting to dislike looking at all this hardware on towers, and high lease costs associated with all of the antennas on each tower, as well as tower loading consideration, most operators today use *duplexing*, which allows a transmit and receive path to share the same antenna and cable. This means the four antennas needed before can be reduced to two antennas. In fact, some antennas today build in receive diversity, such that each sector can actually only have one antenna, yet have two paths.

Many of the newer base stations build in a duplexer so that they have only two ports per sector: a duplexed transmit/receive port and the diversity receive port.

As mentioned, sites generally have two receive antennas on each sector. By placing these diversity receive antennas at different horizontal positions, they have what is called *space diversity* and will thus have a different fading envelope. When the signals from these two diversities combine, the result is a reduction in the amount of overall fading.

While base station hardware is usually protected from the weather by some form of a shelter, the antenna system is usually mounted on a tower and subject to all of the elements, from rain and snow to wind and heat. Hence, antenna systems need to be checked regularly to ensure proper operation. Bad connections, moisture in the system, and general cable degradations are very common, and a periodic maintenance program specifically for antennas needs to be established. Specific testing of base station antennas will be discussed later.

Mobile station antennas are an important concern as well. Naturally, by improving the performance of the mobile's antenna, the performance of the overall system can be improved. While most mobiles will use an omnidirectional single antenna for both transmitting and receiving, some automobile mounted antennas also use space diversity on the receiver to improve reception. As the cellular world moves towards handheld lower power mobiles, improving the antenna systems can improve overall quality.

2.9 Improving Performance

While this book is not intended for performance or design engineers, the base station as well as the mobile station technician would be well served to have an understanding of performance issues related more to network design then maintenance of the hardware. Part of a technician's role is to differentiate network design problems from hardware performance. This section will provide a brief overview of several of the general problems with cellular systems and some of the ways they can be overcome. The individual technology chapters will get more involved on many of these issues, and if more information is desired, most of today's engineers rely on the works designed specifically to cover these issues.

When we speak of system performance, a term most technicians will hear rather often is the *Erlang*. The Erlang, named for Swedish engineer A. K. Erlang, allows performance engineers to have an objective measurement of traffic volume. Just because you have 100 mobile phone subscribers in an area does not mean you need to have 100 open channels at all times. Most people do not use their phones 100% of the time. If a person uses his or her phone 100% of a set amount of time, it can be represented as 1 Erlang; if they used it half the time, they would generate 0.5 Erlang. Thus, if our 100 subscribers were to average using their phones 10% of the time, we would say they generate 10E.

In the high offices of cellular carriers, the term *blocked calls* is often thrown about. Blocked calls indicate the amount (usually in a percentage) of time a user will not be able to get a call out because of network capacity. It is also known as the grade of service. Generally, performance engineers will work off a chart showing channels required versus Erlangs in order to compute how many channels are needed to keep the number of blocked calls, or the grade of service, above an acceptable level—perhaps 2% for a cellular operator.

Today's cellular design and optimization engineers have a variety of state-of-the-art tools to use to ensure their network is at peak performance. Drive-test tools, which can measure and record information on signal strength and quality, are integrated into sophisticated mapping software packages. Advanced software packages even allow an engineer to play with the network coverage settings, such as the effect of raising or tilting an antenna, adding base stations, or perhaps changing the output powers.

One of the primary issues for performance engineers is getting adequate coverage geographically. Holes in coverage are inevitable, but by changing antenna angles, adjusting power levels, and changing antenna heights, all while keeping in mind capacity issues, an engineer can limit the amount of dropped and poor quality calls substantially.

In line with this is a second major issue for performance engineers: too much coverage. The issue of cochannel interference is a problem for most network designers. Because of the increase in capacity and limited bandwidth, planners need to reuse frequencies more often than in the past. This leads to cochannel interference issues, where two signals on the same frequency find their way to the wrong terminal, thus causing interference and sometimes a crossover condition (where one would listen to the wrong conversation). Planners must be prudent in their frequency channel selection, as well as with other tools at their disposal, such as the *supervisory audio tone* (SAT) and digital color code (more on these in Chapter 3). Each base station will have a SAT and digital color code assigned to it, and this allows a mobile station to ensure it listens only to the base station to which it has been assigned.

CDMA uses the same frequency for all channels; thus, cochannel interference is assumed. While this will be discussed further later, CDMA uses PN offsets to distinguish sectors and base stations. The PN offset of one sector will be noise to a mobile listening to another PN offset. Too much overlap of power between multiple offsets is called *PN pollution*, and this can cause serious quality problems.

2.10 Squeezing Out As Much As We Can—Repeaters and Front Ends

As mentioned earlier, holes in coverage are inevitable (see Figure 2.15). This is generally due to terrain or manmade structures that will limit the ability of the RF signal to propagate. For instance, the average shopping mall does not have a whole lot of windows, and RF just does not propagate well through concrete. A very large mall is a prime place for people to use cellular phones, however, and the goal of the carrier is to get people using their handsets.

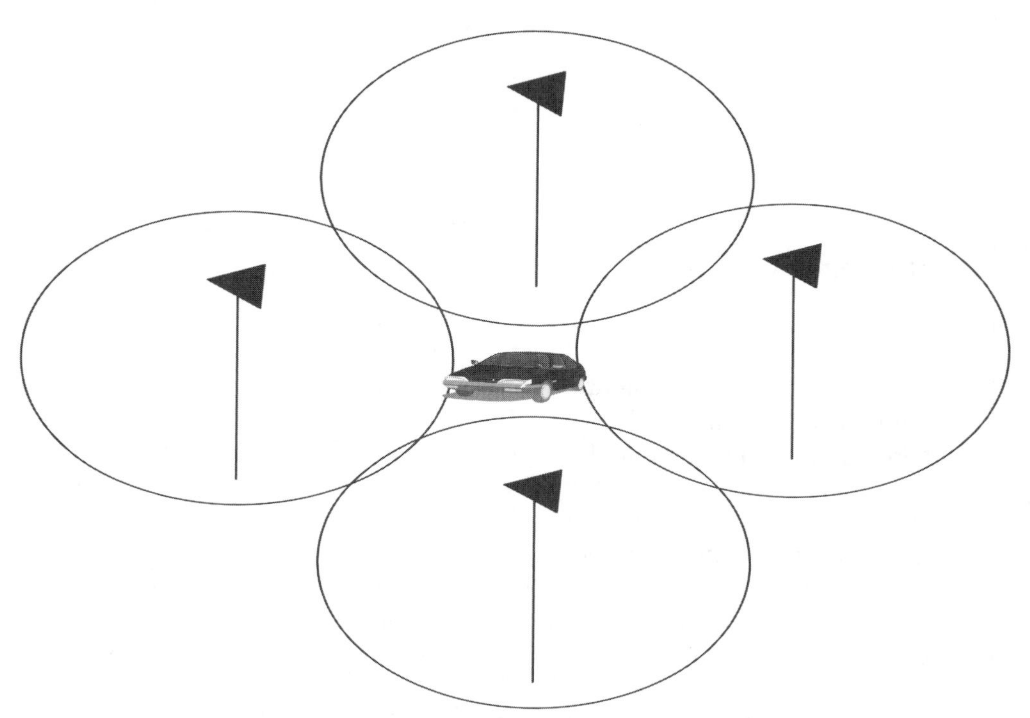

Figure 2.15 Holes in coverage can be dealt with in a variety of ways.

There are several ways carriers can solve these hole issues. The most obvious is to install a new base station. This is quite costly, however, and when we consider the backhaul (the connection to the network), the base station hardware, the maintenance, and the space needed, it usually doesn't make sense unless the carrier can be sure it will stay at a high level of capacity. This is usually not the case when it comes to holes. Many manufacturers now have micro and pico cells. These very small base stations, usually with limited capacity, can be hidden in a mall very effectively and installed very easily, and they require little if any maintenance. For shopping centers, businesses, and airports, microcells and picocells are often a good alternative, as they increase not only coverage but also capacity.

For reaching out and increasing coverage areas, like many rural carriers need to do, the most obvious solutions are to increase the height of the antenna. A general rule of thumb, called the *6 dB/oct rule*, states that by doubling the antenna height, you get a 6-dB gain increase. Of course, this assumes generally flat terrain. Better performing and placed antennas will also improve reception.

Another route carriers can take is the repeater. The repeater is essentially a low-noise amplifier. By placing the repeater in a hole, the overall signal in both directions will become magnified, allowing better quality service (see Figure 2.16).

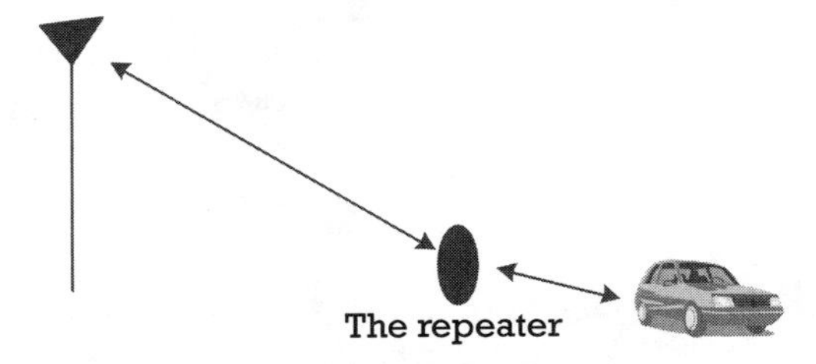

Figure 2.16 A repeater can extend coverage by amplifying signals.

However, anytime you add a nonlinear device in your network, you risk adding interferers (via intermodulation). When you use an amplifier, it increases not only your desired signals, but the noise floor as well. Also, because repeaters simply expand the coverage area of the already existing sector, they do not increase capacity at all. However, at a very low cost for implementation, and a very good return on the investment, repeaters have been used extensively in AMPS to increase coverage gaps.

In rural areas, carriers have little capacity but very large geographical areas to cover. In the past, most rural carriers sold only bag phones, many capable of 4-W outputs. Today, more and more rural carriers are selling handheld phones, most with a maximum output of 600 mW. This magnifies their already existing coverage issues. Thus, the concept of the tower-top amplifier was developed.

Like the repeater, the tower-top amplifier is a low-noise amplifier designed to allow the base station to receive lower level signals. One of the issues with tower-top amplifiers is that they very often will add noise as they increase the desired signal. Known as the noise figure, this can be the limiting factor for their ability to receive low-level signals. The ability of a system to receive a signal and demodulate it correctly is very often determined by the S/N ratio.

In response to this issue, many leading-edge companies have used supercooled superconductor technology to create very-low-noise amplifier and filter systems that can dramatically increase coverage and performance. These new front ends are expected to dominate this market as the cost comes down. If you can lower the front-end noise of your receiver, while your received signal remains the same, you will increase your S/N ratio, which will allow you to receive signals from further away.

2.11 Systems in Use Around the Globe

While wireless technology will continue to evolve and change, it may be useful to have a brief overview of where the various formats are in use around the world. Naturally, this may change even by the time this book is published.

In Europe, the issue is very clear. Apart from very few remaining analog networks (TACS and NMT), GSM has taken over. Europe uses the 900-MHz band for GSM cellular and has also allocated the 1,800-MHz band for what is known as Digital Cellular System 1800, or just DCS-1800 (similar to PCS in the United States).

GSM was designed specifically for the European market, which meant the main focus was the ability to roam from country to country and, hence, network to network. To meet this goal, GSM has a very open standard for the air interface, and the network elements are clearly specified by the European governing body, *European Telecommunications Standards Institute* (ETSI). *General Packet Radio Service* (GPRS) is a data format designed to integrate seamlessly to the GSM network. Most operators in Europe today offer GPRS.

Europe also has installed a lower tier "cordless" format known as DECT. DECT is designed to act as a step forward from a traditional cordless phone by allowing the user to make and receive calls from a variety of public locations.

Most European nations have allocated and some have auctioned off new spectrum specifically for 3G networks. WCDMA will surely be the format of choice for these networks. It is assumed that eventually WCDMA may replace GSM, but given the economies of scale and widespread deployment coverage that GSM enjoys currently, this would seem a task even larger than replacing the AMPS networks was in the early days of digital cellular.

In Japan, the situation is also somewhat clear. PDC, a digital format very similar to NA-TDMA, is the primary cellular format; however, CDMA is deployed with a major Japanese operator, and offers some advantages over the PDC network in terms of capacity and technology road map.

Japan had a love affair with the digital mobile phone system early and capacity grew very quickly. In response, a lower tier cordless format was instituted, more as a WLL system. PHS was launched in 1995 at much lower rates than its big brother, PDC. For pedestrian, home, and business uses, PHS proved to be a cross between a cordless and a cellular system, generating almost 4 million subscribers in the first year.

Japan, as the leader in cellular deployment, deployed the world's first true 3G network, with NTT DoCoMo's WCDMA network. This network, in its own frequency band, has been slow to add subscribers, but its feature-rich capabilities are sure to catch on as the economies recover and demand for Internet access increases.

The rest of Asia and Africa is not nearly as clear cut as Japan and Europe. Although most of the developing countries in Africa seem to lean towards GSM, CDMA has many strongholds. In South Korea, the largest (as well as one of the first) networks, CDMA, currently operates, but there is talk of a WCDMA deployment for 3G. China is predominantly GSM but has an increasing CDMA presence, which may grow even larger than GSM. In addition, China is developing and deploying *time duplexed-synchronized code division multiple access* (TD-SCDMA), a variation of WCDMA.

The Americas are where the real arguments between the technologies take place. Most of the Americas had a substantial investment in the analog system (AMPS), so it made sense to design the next generation to fit into the current structure, without having to replace the AMPS infrastructure or find a new frequency band. Thus, NA-TDMA, or IS-54 and later IS-136, was born. IS-136 uses the same channel spacing as the analog system and can use the same control channels as well. This allowed for a seamless transition to digital.

In all of North and South America IS-136 had become a leading force, largely because of the ease of implementation. With three time-domain channels per 30-kHz channel, this system can quickly improve system capacity over AMPS.

The Americas did take another path as well in designing a system for capacity. CDMA was designed not for ease of implementation (although it was a consideration), nor for ease of roaming, but rather for capacity. This concept was accepted by many carriers early on in PCS as well as cellular in the United States, Canada, and later throughout Latin America.

GSM is also very prevalent in the Americas, although originally only in the PCS bands. This is because, like CDMA, it has a wider occupied bandwidth (200 kHz), but it does not provide a large-scale capacity increase versus AMPS (200 kHz per channel plus guard bands takes up at least seven analog channels and yields eight digital traffic time slots; though, of course, this oversimplifies capacity planning) when compared to cdmaOne. Also, GSM did not have the ease of integration that NA-TDMA gives a carrier. GSM took off early on in the United States and seemed destined to take over the PCS market. In fact, the very first PCS system, Sprint Spectrum in the Washington, D.C., area, seemed to be an early indicator and provides excellent service. Many look to the technical reasons for reasons why GSM did not instantly dominate the American market as it did much of the world, and some would simply blame the "homer" syndrome—Americans prefer a home-grown technology.

GSM 1900 was unable to establish nationwide coverage in the United States early on. One of GSM's main features is the ease of roaming. Without nationwide GSM coverage, which requires working relationships between carriers, GSM faces a hurdle. In the United States, AT&T has a large portion of the country covered with its IS-136 system, and Sprint has CDMA coverage in most major cities (and by using dual-mode handsets, they have analog roaming agreements in most of the rest).

GSM offered carriers and customers more services and features immediately, but many new carriers expected CDMA to leap ahead in this area and provide an easier integration to a third generation format. To help push the GSM cause in North America, a powerful marketing organization has been put together and has made great strides in the United States. Many of the GSM operators began consolidating. Eventually, they were brought under the T-Mobile brand, which has become a global GSM network, and is a national carrier in the United States.

As stated earlier, CDMA and TDMA are both dual-mode standards (meaning they allow for digital as well as analog operation) and therefore often utilize dual-mode handsets. This allows much easier implementation of nationwide roaming agreements and coverage, as CDMA carriers can offer their customers roaming services even in areas without digital coverage. As digital networks expand, however, reliance on AMPS has become very limited.

NA-TDMA operators in the late 1990s and early 2000s began to realize that they did not have a clear roadmap to high-speed data. While several proposals could integrate, NA-TDMA would never have the economy of scale of GSM, a worldwide format. Thus, deploying new infrastructure to support any upgrade would be much more costly than deploying GSM infrastructure. Although this obviously oversimplifies the decision process, it essentially is the reason most of the NA-TDMA operators in the America's began deploying GSM, first in their PCS bands and then in their cellular bands. Eventually, most plan to phase out their NA-TDMA in favor of GSM and take advantage of the road map GSM offers, including GPRS data, EDGE, and eventually WCDMA.

Thus, today's worldwide digital cellular and PCS market is dominated by GSM, with CDMA still growing and NA-TDMA and PDC appearing to have peaked. WCDMA and CDMA2000 appear to be the dominant 3G formats for the future, with TD-SCDMA being used in China to a limited effect.

In analog, while TACS and NMT are gone in Europe, AMPS in the Americas doesn't seem to be going anywhere for at least another few more years, although it would seem to be rapidly disappearing in importance as digital networks complete their footprint.

2.12 Conclusion

The evolution of wireless networks from analog to digital systems has certainly been the major issue in the wireless community for the last several years. Going over some of the history behind the industry and some of the basics should give you a good understanding of where we have come from and why, which should make a discussion of where we are currently with wireless systems a bit easier to convey.

CHAPTER 3

Advanced Mobile Phone Service

3.1 The Basics of AMPS

Perhaps the hardest part of understanding the technical aspects of the AMPS system, and similarly any wireless format, is the endless parade of industry-unique terms and abbreviations. This chapter will explore the AMPS system and hopefully explain most of the terms and abbreviations. Many of these terms and concepts are common across all of the formats, so this section serves as a basis for all of the standards. Additionally, NA-TDMA and cdmaOne are dual-mode standards, indicating that they include the AMPS system in addition to their digital standards. This chapter examines the original basic AMPS system. Both the current NA-TDMA standard, IS-136, and the cellular band cdmaOne standard, IS-95, include some changes to the AMPS implementation, such as adding authentication procedures and special services, such as caller ID. Those features and enhancements will be discussed in detail in the digital format chapters.

3.2 The Analog Network

Many of the terms used to describe the network architecture in the AMPS standard are different from both those used in the later digital systems as well as the terms used today on the street. For instance, the AMPS standard refers to base stations, or BTSs, or cell sites, as land stations. Handsets or mobiles are known as mobile stations. The cellular switching center is called the MTSO (see Figure 3.1).

As mentioned in Chapter 2, the AMPS standard does not specify the interfaces between base stations and the switches to the outside. This means that systems will differ from manufacturer to manufacturer. Until the mid-1990s, communication between different manufacturer's systems (such as what is needed for roaming) was difficult and complicated because of this. In the 1990s, however, IS-41, a standard that covered this type of communication, was developed and slowly implemented.

Communications between base stations and the switch also vary. In AMPS systems, this connection is called the landline—despite the possibility that the connection could be a microwave radio link or perhaps a leased T1 line.

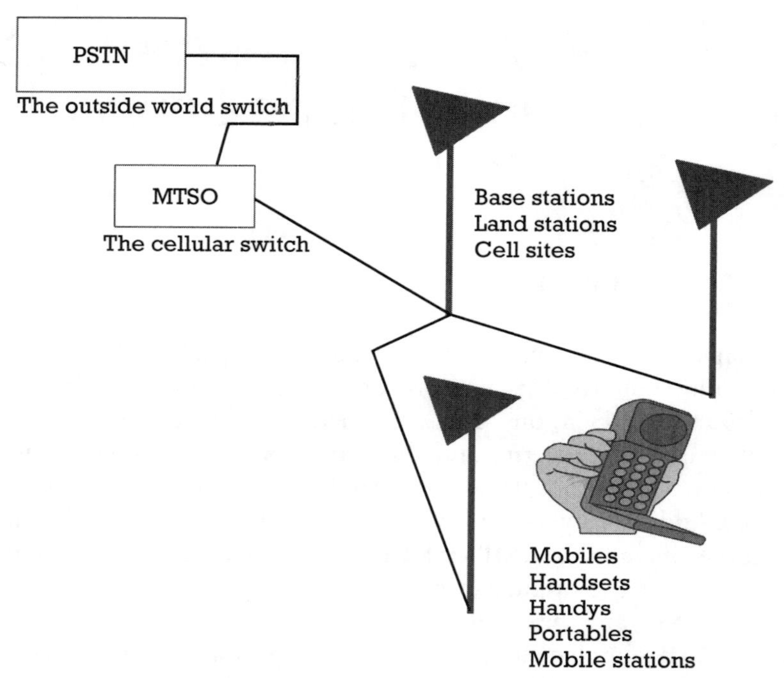

Figure 3.1 Different names for the various pieces of AMPS network architecture.

3.3 Signaling on the Analog System

AMPS is an analog system, which means that the voice is transmitted using FM as described in Chapter 1. This does not mean, however, that digital information does not need to be transmitted from the base station to the mobile and from the mobile to the base station.

AMPS performs this transmission of digital sequences by using FSK. Specifically, AMPS uses Manchester-coded binary shift keying at 10 Kbps. Ones and zeros are represented by transitions of the sine wave from either 8 kHz above or below the carrier frequency (see Figure 3.2). From the data rate, you can convert to see one bit occurring every 100 ms. In this system, a one is represented by a transition from –8 kHz to +8 kHz, precisely in the middle of the 100-ms interval, while a zero is represented by a transition from +8 kHz to –8 kHz.

Another method of transferring information is to use SATs and the supervisory tone. A SAT is one of three frequencies (0:5,970 Hz, 1:6,000 Hz, 2:6,030 Hz) used throughout a conversation on the forward and reverse voice channels, as a means of identifying that the base station is talking to the intended phone and as a means of signaling between the phone and the base station during a call. The base station assigns one of these SAT frequencies to each of its voice radios. When a user accepts

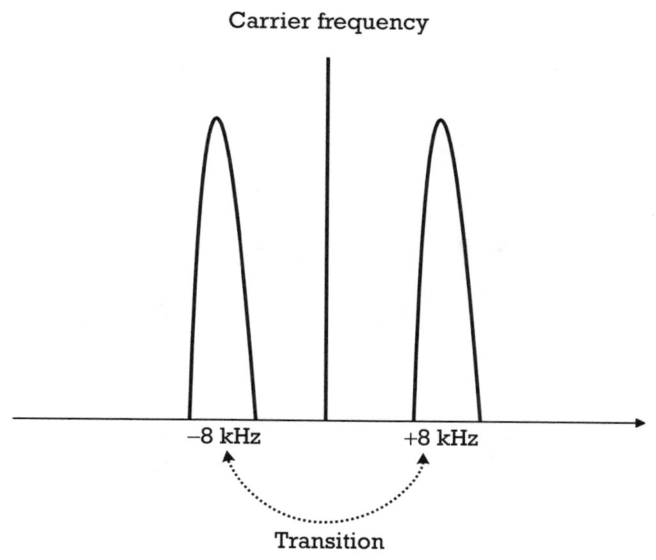

Figure 3.2 Digital ones and zeros are represented by shifts in the frequency of ± 8 kHz.

or originates a call, the base station tells the mobile which *forward voice channel* (FVC) to use as well as which SAT to expect, known as the *supervisory audio tone color code* (SCC) in the mobile station control message. The base station then begins transmitting the specific SAT on that FVC. The handset then moves to that channel, listening for the correct SAT. If all goes well the handset will detect, filter, and modulate with the same tone on the *reverse voice channel* (RVC). This then completes the loop, and the network can be assured that the user is on the correct radio. Not only did the handset have to move to the correct frequency channel, but it also has to find and retransmit the correct SAT (see Figure 3.3).

SATs are also used after a call is established as a means of signaling the occurrence of events such as confirming orders and flash requests, and also as a means of ensuring the integrity of the link between the base station and the handset. This is done in cooperation with the signaling tone. The SAT is also very important during handoffs, where the handset and base station can use the closing of the SAT loop to ensure that the handoff did indeed take place correctly.

In FDMA systems such as AMPS, different users are separated by different frequencies. A frequency reuse system is employed (i.e., the use of cells with different frequency groups) to increase spectrum efficiency and usually requires a significant planning effort to optimize. Even with an optimally designed reuse pattern, some cochannel interference will probably occur. This then plays into the first reason for the SAT, to ensure that the user is using the right frequency on the right base station. A phone that is on the same frequency but transmitting on a different base station

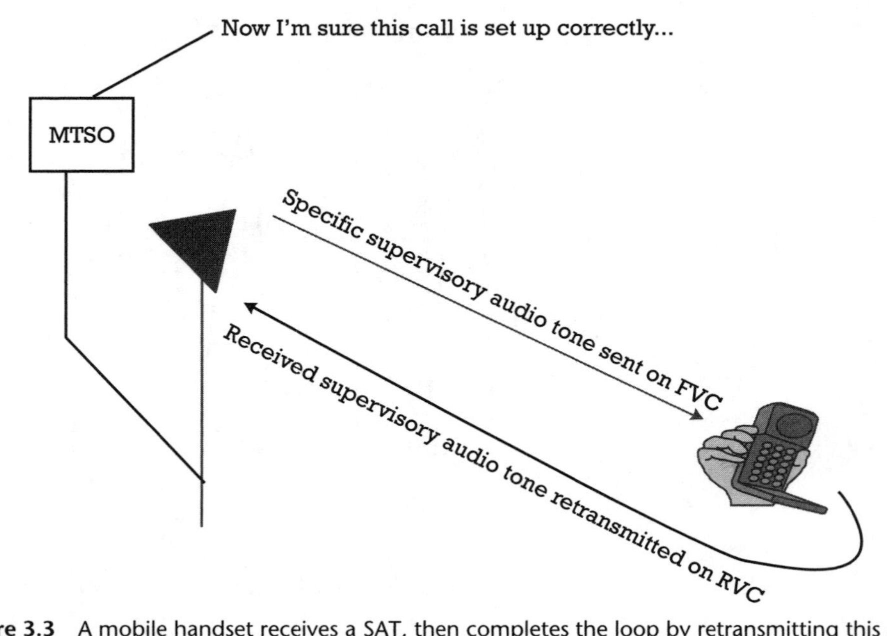
Now I'm sure this call is set up correctly...
MTSO
Specific supervisory audio tone sent on FVC
Received supervisory audio tone retransmitted on RVC

Figure 3.3 A mobile handset receives a SAT, then completes the loop by retransmitting this same SAT. This ensures proper call setup.

would have a different SAT frequency assigned, and the base station would then be able to reject the signal from the interfering call. Because the SATs must be detected and modulated within fairly exacting specifications (SAT frequency of the base station must be within ±1 Hz for the mobile—see Table 3.1), this process ensures that both the base station's voice radio and the handset are operating correctly and will be able to provide quality service to the customer during the call.

In the event that a SAT is not detected, or the SAT frequency does not match what would be expected, the *fade timing status* is enabled. This clock counts 5 seconds. If no SAT has been detected by this time, it turns off the transmitter, assuming that the call was lost to a fade.

Table 3.1 Determining SAT Frequencies

Measured Frequency	*Measured SAT Determination*	*Where*
$f \le f1$	No valid SAT	$f1 = 5{,}955 \pm 5$ Hz
$f1 \le f < f2$	SAT = 5,970	$f2 = 5{,}985 \pm 5$ Hz
$f2 \le f < f3$	SAT = 6,000	$f3 = 6{,}015 \pm 5$ Hz
$f3 \le f < f4$	SAT = 6,030	$f4 = 6{,}045 \pm 5$ Hz
$f4 \le f$	No valid SAT	
No SAT	No valid SAT	

The *signaling tone* (ST) is the second type of tone used in AMPS. The tone is generated by the handset, as opposed to the SAT, which is originated by the base station. The ST is a 10-kHz tone that is frequency modulated on the carrier. The ST is used in combination with the SAT throughout the phone call. The on or off states of the SAT and ST, as well as the lengths of the ST, indicate particular occurrences. For instance, with the SAT and ST on for 1.8 seconds, a release of the call is indicated to the network.

The mobile can also confirm various orders with the SAT/ST, using the same methods. For instance, the handset can confirm an order to perform a handoff by sending the ST and the SAT for 50 ms. A variety of other sequences indicate particular alerting, acknowledgment, and conversation states using the SAT and the ST. The most common is probably the hookflash, which is the SAT and ST sent for 400 ms, followed by a transition to SAT only.

3.4 Preparing Analog Signals

As far as transmitting the analog voice (and the tones), procedures are used to ensure the process is performed correctly. Specifically, five processes are performed before the signal is frequency modulated, amplified, and transmitted (see Figure 3.4).

The first process on the transmit side is called compression, the opposite of which on the receive side is called expanding. Analog voice signals will generally have a wide range of amplitudes or, in other words, a high dynamic range. For instance, the average voice conversation might have a range of 12 dB (which, you will recall, would indicate a difference of 1,600%). The problem with transmitting a high dynamic range is that the weak sounds become very susceptible to noise and the very strong sounds also get distorted.

AMPS uses what is called a 2:1 syllabic compander. This simply means that for every 2-dB change of the signal at the input, the output will only change 1 dB. This compressed signal is than sent to a preemphasis filter.

The preemphasis filter is designed to improve the sound quality of the voice signal. It is used in conjunction with a de-emphasis filter on the receive side. They work together to increase the level of the higher frequency sounds for transmission and then return them to the original level at the receiver.

After this preemphasis, the signal is sent to a limiter and lowpass filter. The purpose of amplitude limiting is to ensure that the FM signal does not exceed 12 kHz on either side of the carrier. The lowpass filter attenuates frequencies above 3 kHz. You will recall the channel bandwidth of AMPS is 30 kHz, so any energy present greater than 15 kHz from the carrier will cause a condition known as adjacent channel interference. Specifically, any energy that would occur at a frequency greater than 15 kHz from the carrier needs to be attenuated more than 28 dB (see Figure 3.5).

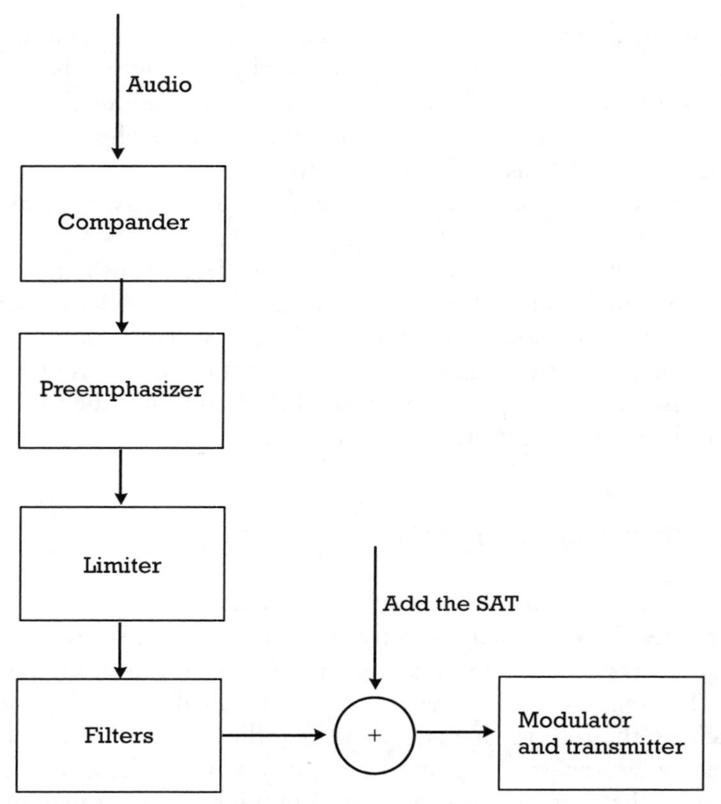

Figure 3.4 Audio signals need to be prepared before they are modulated and transmitted to ensure good dynamic range and prevent interference.

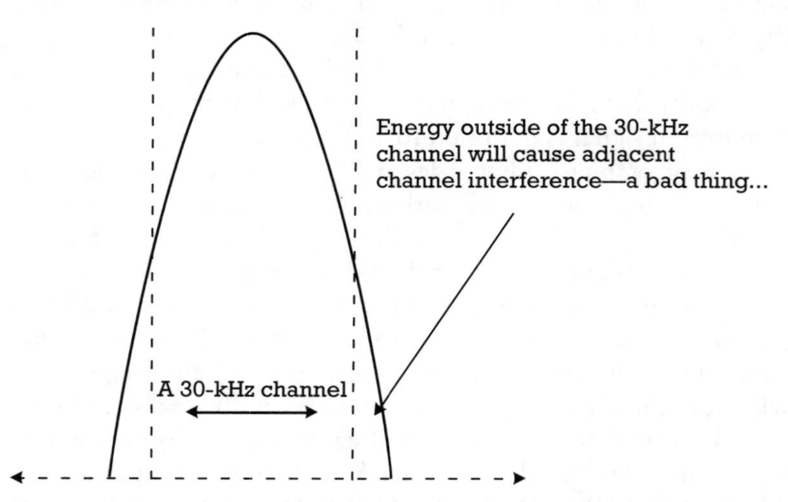

Figure 3.5 Any energy occurring outside the 30-kHz bandwidth of the channel will interfere with adjacent channels.

The last process before modulating, amplifying, and transmitting is to add the SAT to the signal, to ensure the voice channel is set up correctly (as described in Chapter 2).

3.5 Identifying the Mobiles and Base Stations

The mobiles and base stations need methods of identifying themselves to each other as well as the mobile switch (see Figure 3.6). Every phone in a system will be assigned a standard 10-digit telephone number, in this case called the *mobile identification number* (MIN). The number takes the same format as landline telephone numbers, with an area code identifying the mobile station's home service area, a three-digit exchange number, and a four-digit subscriber number.

The mobile also has an *electronic serial number* (ESN). The manufacturer permanently assigns this 32-bit code. The ESN consists of three fields, including an 8-bit manufacturer code, an 18-bit unique serial number, and 6 bits that are reserved for later use. Usually, the phone manufacturer implements a security system that renders the phone inoperative if any attempt is made to change the ESN. If false ESNs can be entered, dishonest people can easily commit fraud against the carrier (the cloning of phones).

The mobile station also has an identifier that tells the network what capabilities the mobile station has (see Figure 3.7). For instance, some early mobiles were designed before the extra spectrum, and therefore extra channels, were assigned. You should remember from Chapter 2 that originally there were only 666 channels

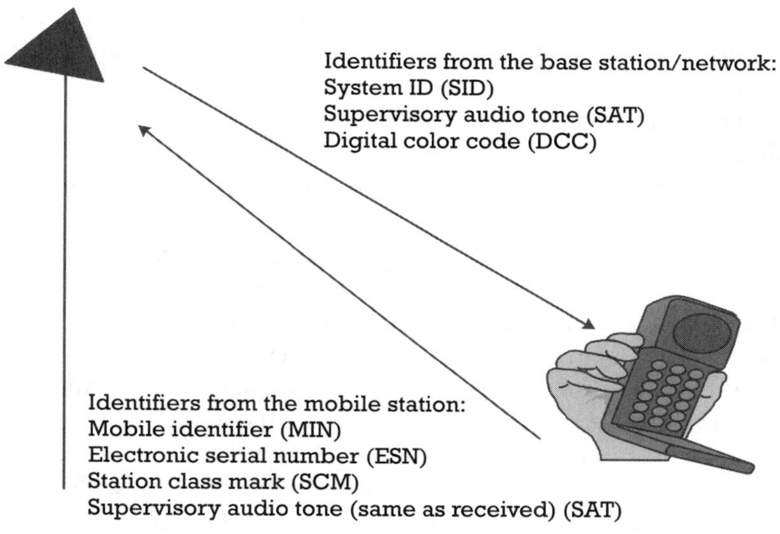

Figure 3.6 There are a variety of identifiers for the base station as well as the mobile station.

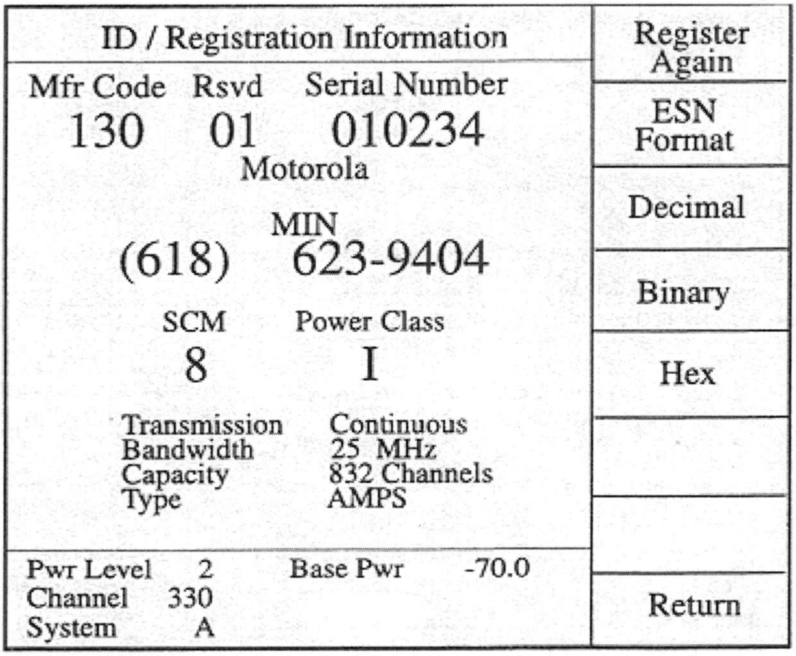

ID / Registration Information	Register Again
Mfr Code Rsvd Serial Number 130 01 010234 Motorola MIN (618) 623-9404 SCM Power Class 8 I Transmission Continuous Bandwidth 25 MHz Capacity 832 Channels Type AMPS	ESN Format Decimal Binary Hex
Pwr Level 2 Base Pwr -70.0 Channel 330 System A	Return

Figure 3.7 The base station (or in this case, the communications test set) can interpret the various identifiers and gain a substantial amount of information about the phone.

before the FCC added the extra spectrum containing 166 more channels, for a grand total of 832 channels. The network needs to know whether the phone can access all of the channels; otherwise, it might tell it to hand off to a voice channel that it could not possibly tune to. Also, mobiles have different maximum power levels of between 600 mW and 4W. This identification is called the *station class mark* (SCM).

The mobile phone also needs to be programmed with its preferences for acquiring service. For instance, the phone needs to be set up for the A band or the B band. Also, preferred paging channels can be identified to speed access to the home system (the FIRSTCHP parameter).

The base stations and the networks also share identifying codes with the mobiles as the mobile station attempts to get service. Every system has its own identifier. This *system identifier* (SID) is assigned by the FCC and designates a carrier in a specific geographic area. The SID, and a subset called the *network identifier* (NID), is also programmed into the mobile stations when the user subscribes to the service. This can then be compared to the received SID in order to let the phone know that it is either in its home system or roaming. A roaming indicator can then come on to tell the user that he or she is probably paying a bit more to use the phone in this area. In addition, the A band carriers will be assigned odd numbered SIDs, while the B band carrier will have even SIDs.

As mentioned earlier, base stations also transmit a SAT and a *digital color code* (DCC), which are used to identify specific base stations. The SAT is sent on the voice channels, while the DCC is sent on the control channels (in digital form).

3.6 Frequency Channels

As discussed in Chapter 2, AMPS uses the cellular frequency band and is described as an FDMA system using FDD. AMPS channels are 30 kHz wide, and because it is an FDD system, one voice channel actually uses two 30-kHz channels. The uplink or reverse link (from the mobile to the base station) uses the lower frequencies (if you remember, lower frequencies propagate better in most situations), with the downlink or forward channel using the frequencies 45 MHz higher.

AMPS channel numbers can easily be computed to frequencies (which is a task the mobile must perform) by using a simple formula. For the uplink or reverse channels:

$$\begin{aligned} &\text{If } C \le 799, \text{ then} && F(C)\text{kHz} = 825{,}000\,\text{kHz} + 30(C)\text{kHz} \\ &\text{If } C \le 991 \ge 1{,}023, \text{ then} && F(C)\text{kHz} = 825{,}000\,\text{kHz} + 30(C - 1{,}023) \end{aligned} \tag{3.1}$$

where

C = channel number

F = frequency

The downlink frequency is simply the reverse channel plus 45 MHz.

The reason for the skipping of channel numbers to 991 for the last 33 channels pertains to the type of mathematics digital systems use to perform computations (in this case, it is called *two's complement arithmetic*), and setting up the numbering scheme this way allows for easier computation in the mobile.

Thus, the cellular industry has 832 channels to work with. In the United States, the FCC has given out 1,466 licenses, two in each market. Of these 1,466 licenses, two carriers are awarded licenses in 428 *rural service areas* (RSAs) and 305 *metropolitan statistical areas* (MSAs). Thus, in each market, two carriers share the 832 channels, 416 each.

3.7 Control and Voice Channels

There are two general types of channels designated in an AMPS system: control channels and traffic channels. Control channels contain the digital information (ones and zeros) that is needed to establish the mobile station on the network (registration), to set up calls from the mobile station (called a *mobile origination*), and to set up calls coming in for a particular mobile station (called a *mobile page*). Traffic channels are active during voice conversations, but they also do contain the digital information needed to keep a call up. A mobile station thus tunes to and receives either a control channel or a traffic channel at any given moment.

AMPS designates 42 channels as control channels, 21 for each carrier. They fall in the center of the assigned spectrum, channels 313 to 354. You might note that because cellular systems use a frequency reuse pattern of seven, if each cell has three sectors, there would need to be 21 control channels in order to assure one different channel for each cell in the pattern. The rest of the channels are traffic channels, although some carriers can use additional traffic channels as control channels should a high-capacity issue call for it. In most cases, one primary control channel supplies the overhead messages in each cell. Other terms for control channels include setup channels and paging channels. Although these three terms are usually synonymous, a distinction can be made in that the control channel is the primary carrier of overhead information for initialization, while paging and setup channels can be secondary channels mobile stations move to and monitor in the idle state.

Control or setup channels are also sometimes called paging channels, in the case of the base station or downlink, and access channels in the case of the mobile station or uplink. The paging channels are used to set up calls that originate from the base station, while the access channel is used to set up calls that originate with the mobile. There are other tasks these control channels perform as well, including serving as a method of acknowledging receipt of messages from each other. Generally, the paging channel is designated the *forward control channel* (FOCC) and the access channel the *reverse control channel* (RECC). Data sent on the control channels is block encoded and repeated to ensure correct reception.

Thus, a base station will broadcast information the FOCC to all mobiles who are monitoring it (those not on a call). If the mobile station needs to respond or ask the base station for something, it will have to find a RECC and transmit on that channel. The AMPS standard includes a process for mobile stations to attempt to acquire a RECC called the random access protocol. Thus, the FOCC is one to many, while the RECC is many to one.

3.8 The FOCC

The FOCC is transmitted from the base station continuously. The FOCC is intended to be received not by one mobile, but rather by all of the mobiles in its area. Thus, a mobile will scan the 21 control channels in its band, searching for the control channel with the strongest signal. It can then assume that the acquired control channel is being transmitted from the nearest base station.

Because the FOCC is the entryway for a mobile into the network, any adjustment in the transmission power of the channel can help define how large or small the cell's coverage will be. Performance engineers can also adjust receive thresholds at the base station for the same purpose.

The FOCC is divided into three bit streams, word A, word B, and a busy-idle stream. Mobiles with even MINs will find their information in word A, while mobiles with odd MINs look in word B. The FOCC continuously transmits information in 463-bit frames (see Figure 3.8), which take 46.3 ms each (10 Kbps). In each frame, word A or B is only 28 bits each, and is repeated five times. Thus, for any given mobile, the bit rate would be:

$$28 \text{ bits}/463 \text{ bits} \times 10 \text{ Kbps} = 604.75 \text{ bps} \quad (3.2)$$

Built into this channel are also bits known as busy/idle bits. These busy/idle bits control the mobile station's access to the system, and they occur at a rate of 1 Kbps (or 1 bit every 10 bits). Because many mobiles will have to use the same RECC to talk to the base station, the busy-idle bits will let a mobile know if someone else is already using it. The mobile can then back off and wait to try later. This RECC access protocol is essential because if two mobiles try to talk to the base station at the same time, neither would probably get through. The mobile station will monitor the FOCC that corresponds to the RECC on which it will attempt to access.

The idle-busy bit also serves a confirmation function. The mobile station knows how long it should take after it starts transmitting on the RECC for the idle-busy bit to change from the idle state to the active state. If it changes too early (under 5.6 ms), it knows another mobile jumped in ahead. It can then back off, wait a random amount of time, and try again. If it takes too long (more than 10.4 ms), then the mobile station can assume the base station was not able to receive the information. It again backs off for a random amount of time and tries again.

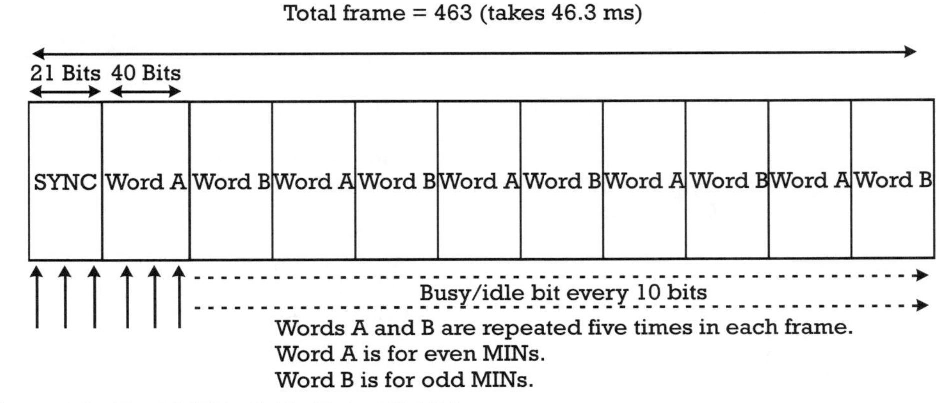

Figure 3.8 The FOCC is divided into 463-bit frames.

System performance engineers can set several variables to affect the way this RECC access protocol performs. The mobile station will count the number of "busy" bits it reads while waiting to use the RECC. If the number of busy bits exceeds a set threshold (the setting message is abbreviated MAXBUSY), the mobile station quits trying. The mobile station also keeps track of the number of times it tries to send the message on the reverse channel but does not get its confirmation. Again, if it exceeds a threshold (MAXSZTR), it abandons the process.

As you can imagine, the base station will transmit a variety of different messages to the mobiles in its cell. Some of the information pertains to all mobiles, while some of the messages pertain to specific mobile stations.

Four messages are transmitted on the FOCC in what is called an overhead message train, called overhead messages (see Figure 3.9). The overhead messages will be different for each base station. These messages include the SYSTEM PARAMETER, GLOBAL ACTION, REGISTRATION IDENT, and the CONTROL FILLER messages. Again, as these messages are intended for all mobile stations, they are called broadcast messages.

The SYSTEM PARAMETER message uses two of the 28-bit words. The information consists of the SID of the base station as well as the number of control channels, both RECCs as well as FOCCs. This message is very important to the mobile stations, as it tells the mobile station if it is roaming or not (and, naturally, this information is quite important to the user, or at least whomever is paying the cellular bill) and it tells the mobile information it needs to use the RECCs.

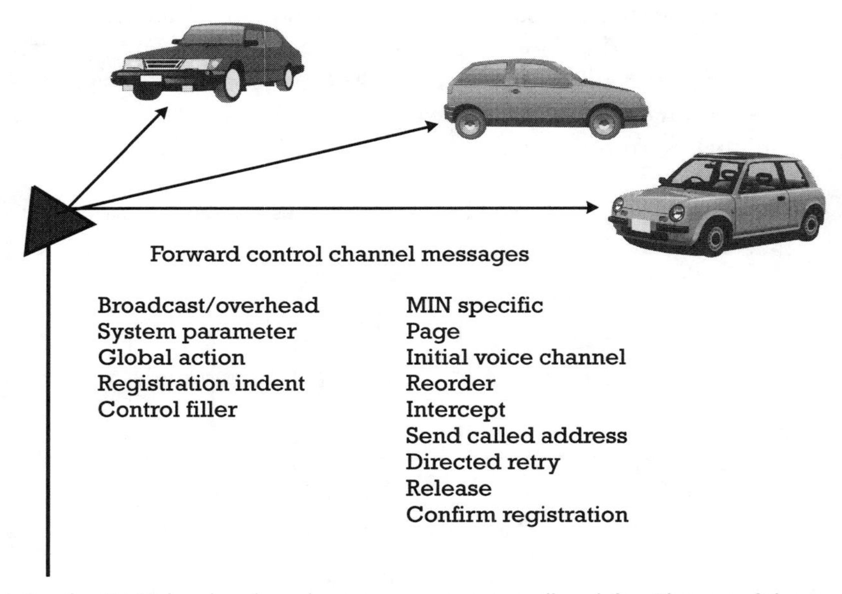

Figure 3.9 The FOCC has four broadcast messages sent to all mobiles. The rest of the messages are intended for specific phones.

GLOBAL ACTION messages contain information that all of the mobiles need to operate in the cell. This includes parameters for setting up the RECC access protocol, which includes the MAXBUSY and MAXSZTR thresholds described earlier.

The REGISTRATION IDENT message tells the mobile stations how often they should perform a registration (i.e., send the base station the pertinent information about the mobile station).

The final broadcast is the CONTROL FILLER message. Because the FOCC broadcasts continuously, there will be gaps between messages. AMPS uses this gap to transmit a message with a parameter called *control mobile attenuation* (CMAC). This parameter tells the mobile stations what power level they should use when transmitting on the RECC.

As mentioned, the rest of the messages transmitted on the FOCC are directed at specific mobile stations. Addressing is based on the MIN of the individual mobile phones. These mobile station–specific control-order messages include:

- *PAGE:* This message informs the mobile station that it has a call coming in and should prepare to set up to receive it.
- *INITIAL VOICE CHANNEL:* This message informs the mobile station of the voice channel it should go to for a call. Included in this message are the parameters CHAN and VMAC, which are the voice channel number and transmit power level.
- *SEND CALLED ADDRESS:* This message informs the mobile station that the base station wants to receive the telephone number that the user wants to call.
- *REORDER:* This message generally is sent when the system has too much congestion to set up a call. The mobile station, upon receipt of this message, will usually give the user a fast busy signal.
- *INTERCEPT:* This message is sent to indicate that the system did not understand the number sequence sent by the mobile station. Again, the mobile responds by giving the user an audible signal.
- *CONFIRM REGISTRATION:* This message lets the mobile station know that the base station received the registration information that it sent on the RECC.
- *DIRECTED RETRY:* This message is used to divide up usage in areas of high congestion. Mobile stations scan the control channels and lock onto the control channel with the highest power (normally the closest). In some cases, where there is particularly heavy traffic, base stations can be placed close together (i.e., using smaller cells). The DIRECTED RETRY message then directs mobiles to different control channels in order to even out the usage. The message gives specific control channels to try and is thus very useful to performance engineers.

- *RELEASE:* This message orders the handset to stop trying to access the system and return to listening to the FOCC.

3.9 The RECC

The RECC uses 48-bit words repeated five times for a 240-bit sequence. Included in this digital sequence is the DCC. The DCC plays a similar role as the SAT on the voice channel (i.e., it ensures it is talking to the right base station, on the right channel, because it is simply copying the DCC it receives from the base station to which it is talking). The RECC only has three messages but, of course, these three messages are crucial to system operation (see Figure 3.10). It is not by coincidence that there are many more messages on the forward channels versus the reverse channels. Essentially, the MTSO (this is where these messages originate) is the primary controller of the performance of the system and, hence, sends the most orders. These three RECC messages are the following:

- *ORIGINATION:* This message is sent when the user presses the send key in order to call someone. Included in the message is the number called, the MIN, the ESN, and the SCM.
- *PAGE RESPONSE:* The mobile station sends this message after it realizes that it has a phone call coming in for it (via the PAGE message on the FOCC). The PAGE RESPONSE message, like the ORIGINATION message, contains the MIN, ESN, and SCM.
- *REGISTRATION:* A REGISTRATION message is sent by the mobile station to the base station before any calls are set up. This allows the MTSO to need

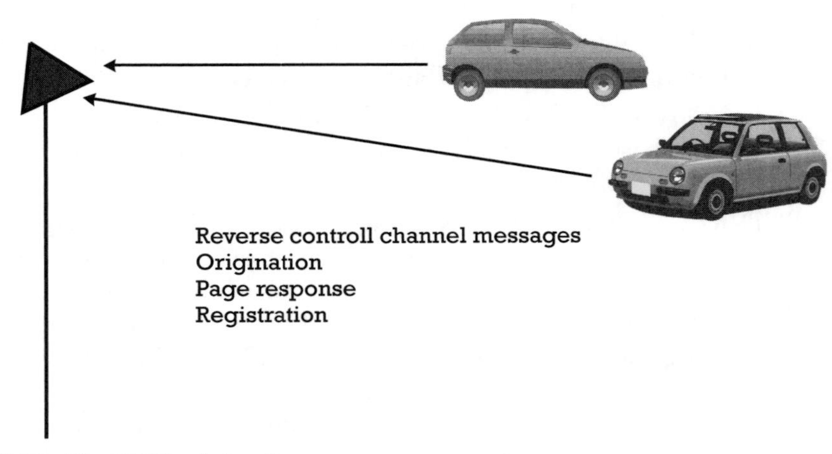

Figure 3.10 The RECC only has three messages.

only to page mobiles in specific areas rather than the entire network. This greatly reduces the resources needed in heavily loaded systems. REGISTRATION messages are essential to set up roaming service. After the mobile station registers in a system different than its home system, the roaming network can inform the home network of the mobile station's location.

3.10 The Voice Channels

After a call is established using the RECC and FOCC, the process switches to the voice channels. These voice channels, the FVC and the RVC, carry the FM voice signals of the conversation. During the course of a voice conversation, however, digital signaling information still needs to be passed from the base station to the mobile station and vice versa. As mentioned earlier, some of the confirmations and indications are sent by the mobile station using the SAT and ST tones. For control orders such as handoff orders, AMPS uses a method called *blank and burst*. This is essentially a digital FSK frame sent for about 100 ms, preempting the voice, but is not audible to the user.

As with the control channels, data sent on the voice channels is coded and repeated, so much so that even with a 10-Kbps transmission rate, the effective data rate on the FVC is only 271 bps, and on the RVC, either 662 bps or 703 bps, depending on whether a one-word or two-word message is sent.

Also similar to the control channels, most of the messages sent are on the forward channels (see Figure 3.11). Changes in the SAT and ST are often used as confirmations. The SAT and ST status is described in the notation: (SAT, ST); for

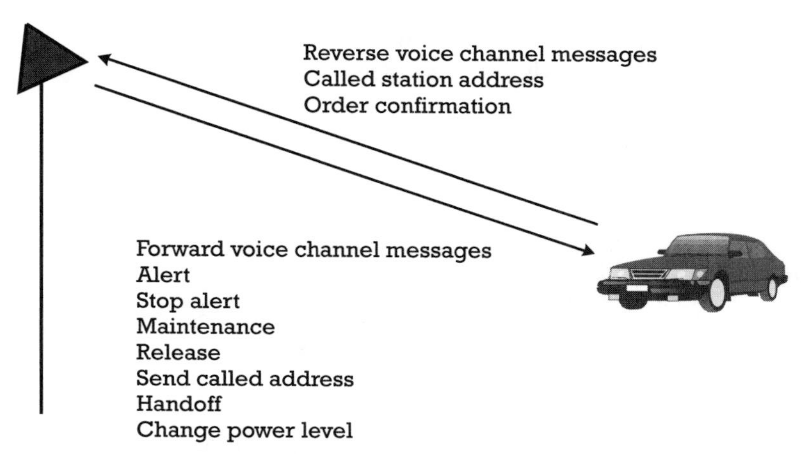

Figure 3.11 Just as with the control channels, the majority of the messages are on the forward channel.

instance, a condition of the SAT on, with the ST off, would be described as a SAT/ST status of (1,0).

Messages sent on the forward voice channel include:

- *ALERT:* If an incoming call is set up on the control channels, the mobile station will move to the assigned voice channel. The FVC then transmits the ALERT message, which causes the mobile station to ring. During this alerting, the mobile station turns on the SAT and ST as a confirmation of receiving the ALERT (SAT,ST status = (1,1)). Once the user presses the send key, the mobile station stops sending the ST (SAT,ST status = (1,0)), which triggers the base station to send the STOP ALERT message, which naturally causes the mobile station to stop ringing.
- *RELEASE:* If the call is ended by the party communicating, this message commands the mobile station to stop using the voice channel and return to monitoring the FOCC. The mobile station confirms reception by changing the SAT,ST status from (1,1) to (1,0) for 500 ms, followed by changing back to (1,1) for 1.8 seconds. At that point, the land station stops transmitting on that channel.
- *SEND CALLED ADDRESS:* This message orders the mobile to respond with the number it is dialing.
- *HANDOFF:* This message instructs the mobile station to perform a handoff, supplying the new channel number, the SAT of the new channel, and the initial power level to use. The mobile station confirms receipt of the message by changing the SAT,ST status from (1,1) to (1,0) for 500 ms, followed by returning to (1,1) for 50 ms. The mobile station then moves to the new channel, sending the new SAT and ST (1,1).
- *CHANGE POWER LEVEL:* This message orders the mobile station to change its power level. The mobile station confirms receipt of this message by sending its own digital sequence on the RVC.
- *MAINTENANCE:* This message causes the mobile station to basically run through the paging process, without alerting the user. It is used to ensure the mobile station is working properly.

On the RVC there are only two messages sent:

- *CALLED STATION ADDRESS:* This is in response to the SEND CALLED MESSAGE order on the FVC, and it contains the number the mobile station is trying to call.
- *ORDER CONFIRMATION:* This message is used to confirm receipt and implementation of important orders, such as a CHANGE POWER LEVEL order.

3.11 Call-Processing States

There are basically four states that an AMPS mobile station will be in at any given moment. During these various states, it will transmit and receive the various messages and tones described earlier, allowing it to perform the various tasks it needs to in order to provide service to the end users. These states include the initialization state, the idle state, the access state, and the conversation state (see Figure 3.12).

The initialization state is entered anytime the phone is turned on, after a conversation ends, or the mobile station loses contact with the base station it was using. The initialization state involves only the control channels.

The handset will be set up as an A-band phone or a B-band phone. This setting is usually determined at the time of activation, but some phones do allow the user to set this, mainly to allow for roaming access. For instance, when a user is outside his or her home A-band system, he or she might prefer to use the B carrier for a number of reasons.

The mobile station will begin by scanning the control channels in its band (either 313.333 for the A band or 334–354 for the B band). In some cases, the

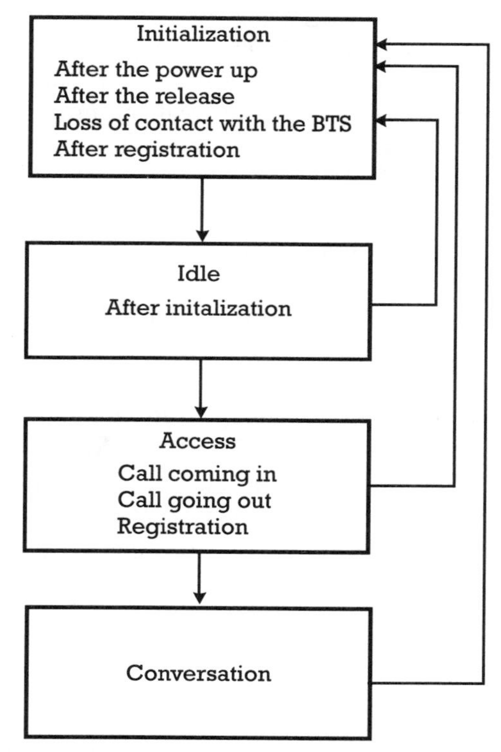

Figure 3.12 There are four states the mobile phone will be in. When in doubt, the phone always returns to the initialization state.

handset can be programmed with preferred control channels to check first. Once the mobile station decides on the control channel with the most power, it begins to read the overhead information from the channel. This includes receiving and processing the four broadcast messages described earlier, the SYSTEM PARAMETER message, the GLOBAL ACTION message, the REGISTRATION IDENT message, and the CONTROL FILLER message.

The initialization process does not require the mobile station to transmit and generally takes 5 to 10 seconds to complete. It is in this process that the mobile station decides whether to display the roaming indicator on the phone if the SID of the system it is on does not match that which is programmed in the phone.

Once the mobile station completes the initialization process, it then enters the idle state. The name of the state is somewhat of a misnomer, for the mobile station is diligently listening to the base station. The mobile station, having moved to the strongest paging channel, monitors the broadcast messages to update call setup parameters and various mobility management tasks.

Several events can cause the mobile station to move to the next state, the access state: the user originates a call by inputting a number and pressing send, the mobile station receives a page message and recognizes its MIN in the message, or the mobile station needs to perform a registration.

During the idle state, the mobile station periodically returns to the initialization state to update its overhead information and rescan for the strongest control channels. As the received power begins to weaken, indicating movement away from the serving base station, the mobile station will begin to rescan the control channels as needed.

The access state is entered whenever the mobile station tries to transmit information back to the base station via the RECC. You will remember the RECC seizure procedure from earlier in this chapter. The various RECC messages already discussed are all sent in this access state, including the REGISTRATION message and ORIGINATION message.

When a call is set up, either via a mobile origination or base station page, the mobile station will move to the conversation state. When the mobile station first moves to the voice channel, it retransmits the correct SAT, serving as the confirmation. Once the base station receives the same SAT that it transmitted, it is sure that the mobile station is correctly tuned and can begin the conversation. During this state, all of the voice channel control messages are sent on either the FVC or the RVC, as previously discussed.

3.12 Transmit Power

One of the most important measurements for technicians is output power, either of the base station or mobile station (and in the case of performance engineers, both).

While AMPS does not have a terribly complex power control system compared to some of the other systems we will examine in this text, there is some dynamic control.

As mentioned earlier, there are a variety of mobile stations, each one's capabilities identified by the SCM. There are three classes of mobile stations (see Table 3.2).

Most mobile stations are either class I, which are mostly car mounted or bag phones, or class III, which encompass most handheld mobile stations.

Class I phones have eight powers levels (0 to 7), where power level 0 is the highest (4W). Class III phones have six power levels (2 to 7), with 600 mW being the maximum.

As you change the power level, the power drops or increases 4 dB, so that every one-step increase in power would equate to 2.5 times more power. The minimum power for all three classes is 6.3 mW.

The base station controls the power levels that the mobile station will use to transmit via the already mentioned messages. While older mobile stations transmit at one power level until ordered to change, some mobile stations are equipped with a feature called *discontinuous transmission* (DTX). This allows the mobile station to use two different power levels, one for when the user is speaking and one when the user is listening (and thus not talking). This naturally increases battery life, not to mention reducing the overall transmitted power in the system, which can reduce some interference. This is a glimpse of variable rate vocoding, which some digital systems utilize.

Base stations generally transmit at a set level. Commonly, a base station will transmit around 20W to 25W per channel, with an antenna height around 100 ft above the terrain. This height of the antenna is sometimes designated *height above average terrain* (HAAT). Of course, this power level and height vary greatly depending on loading conditions and terrain features. Performance engineers regularly manipulate power levels to ensure optimized coverage and performance.

3.13 The Handoff

In the original AMPS system, the handoff is exclusively controlled by the base stations and the MTSO. Essentially, the base stations take *received signal strength*

Table 3.2 Power Classes in AMPS Phones

Power Class	*Maximum Power Level*
I	4W
II	1.6W
III	0.600W

indications (RSSIs) of the designated mobile station and return the measurements to the MTSO. The MTSO then decides, based on specific thresholds, whether to initiate a handoff and whom to handoff to. Typically, AMPS cannot maintain adequate voice quality below –100 dBm, so the threshold is set somewhere around this level, again, determined by performance engineers.

A second threshold is established for the receiving base station, in order to keep the MTSO from performing multiple handoffs as a mobile station moves along a cell boundary or perhaps across terrain that quickly alternates its views of the base stations. The MTSO establishes a threshold for the base station designated to receive the handed off mobile station that is significantly higher than the old base station—perhaps 10 dB higher. This ensures that the mobile station definitely moved out of a boundary area and into the specific coverage area of the new serving cell.

The process of the handoff is performed on the voice channels, using the messages described in the earlier sections (see Figure 3.13). The handoff in AMPS requires a break-before-make situation and is most likely occurring at very low power levels; thus, it is a bit risky and is a cause of dropped calls if not set up correctly.

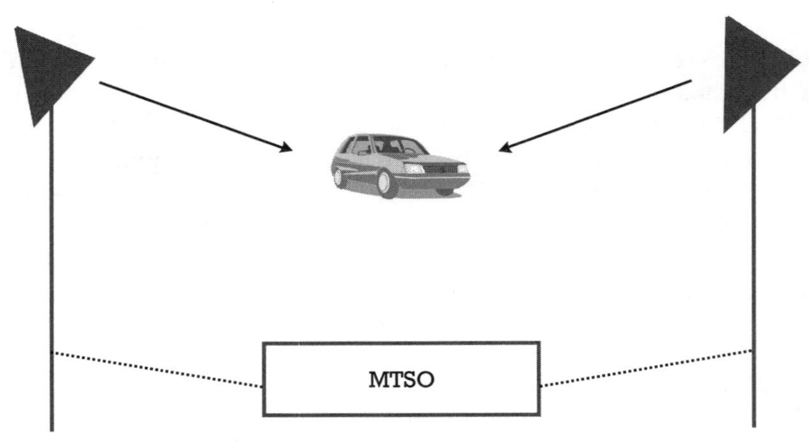

Sequence for a handoff

1. Original base station measures a weak signal.
2. Original base station requests a handoff from the MTSO.
3. MTSO receives measurement results from surrounding base stations.
4. MTSO selects a new base station.
5. Original base station sends the handoff message.
6. Mobile station sends the ST for 50 ms.
7. Mobile station switches to new channel.
8. Mobile station transmits new SAT.
9. New base station detects SAT and confirms to MTSO.

Figure 3.13 The AMPS handoff process—much simpler than what we will see in the next generation systems, but not quite as effective.

3.14 The Problems with AMPS

One of the main issues with the AMPS system is the ability of dishonest people to "clone" mobile phones and thus use other people's minutes on the cellular system for their own use. The primary security protection in AMPS was the MIN and ESN matched pair, but because this information is transmitted on common control channels, it is easily intercepted and copied onto the dishonest person's own phone (see Figure 3.14). While interim security measures have been adopted, including a *personal identification number* (PIN)–type system in which the user punches in a unique code after making a call (it is thought that the RVC is more secure than the RECC), a new security system was definitely needed.

Authentication was included when NA-TDMA and cdmaOne systems were defined. This complex system will be described later in the text, as it is part of the IS-136 and IS-95 standards (remember, AMPS is incorporated in NA-TDMA and cdmaOne—they are dual-mode standards), and the procedure is similar for all formats.

Another reason for the need for a replacement of AMPS is capacity. While an AMPS network can support many ways to increase the capacity, it cannot match some of the capacity improvements had by next generation digital systems. One method for increasing capacity is to simply shrink the size of the cells, thus adding more cells (at lower power levels). There is a limitation due to interference and, perhaps more importantly in today's market, the amount of towers that can be erected in urban areas—for economic (the cost of tower space, installation, and maintenance) as well as environmental (tower moratoriums and regulations) reasons. Increasing the amount of frequencies and thus radios in each base station is another

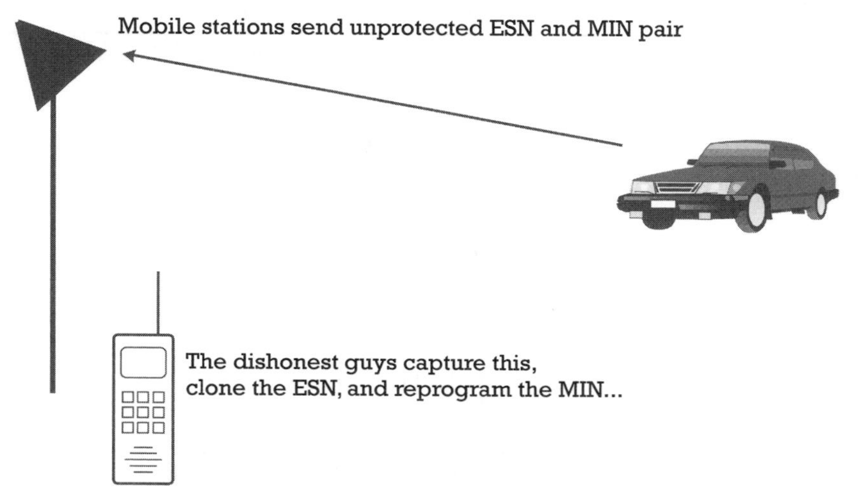
Mobile stations send unprotected ESN and MIN pair
The dishonest guys capture this,
clone the ESN, and reprogram the MIN...

Figure 3.14 All the dishonest guys need to do is monitor the control channels during a registration and they can steal the MIN and ESN pair.

method. Performance engineers can optimize frequency layouts to ensure the most frequency channels get used per sector without an unacceptable amount of cochannel and adjacent channel interference. More spectrum could be allocated. However, the FCC had established a set frequency allocation, and it was not going to add more spectrum for a first generation analog system.

Short of transitioning to a next generation digital system, some carriers did have the option of implementing a Motorola solution called *Narrowband AMPS* (NAMPS). This analog system was very similar to AMPS, except that it used three 10-kHz channels in each 30-kHz AMPS channel. Using a maximum deviation of 5 kHz, instead of 15 kHz, NAMPS theoretically could triple capacity (see Figure 3.15).

In NAMPS, a normal FOCC is used, allowing for dual-mode operation. If the phone is equipped for NAMPS operation, the network can then bring calls up on one of the 10-kHz channels. Although the message formats are the same as in AMPS, NAMPS does not use SATs or STs; instead, it uses a digital code (transmitted in an associated control channel, just like other digital messages).

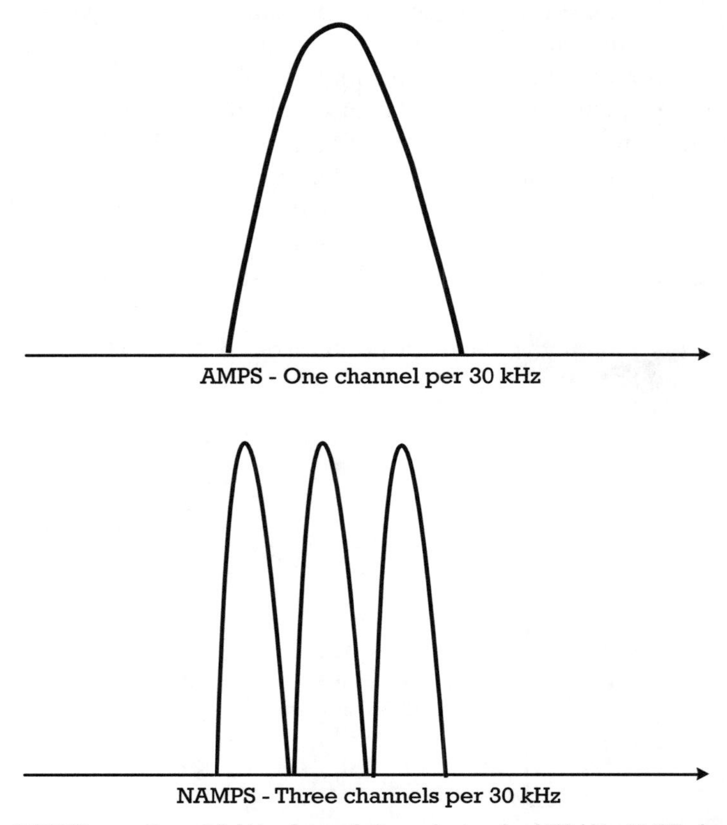

Figure 3.15 NAMPS uses three 10-kHz channels in each standard 30-kHz AMPS channel—thus, a significant capacity increase.

NAMPS was implemented in the early 1990s and incorporated some new features, such as caller ID, voice mail indicators, and short messaging. Some of these features were incorporated into AMPS as well in the dual-mode standards. NAMPS also incorporates early versions of the mobile-assisted handoff. Mobile stations are capable of measuring the bit error rate of the control information on the voice channel, reporting the information back to the MTSO. A handoff can then take place based on this information.

Because NAMPS uses a narrowband signal, it is more susceptible to interference, and thus frequencies cannot be reused as much as in AMPS. This cuts down on the threefold increase in capacity.

Thus, AMPS has been able to stay in place in North and South America by overcoming some of its shortcomings, including security, improved special services, and to some extent capacity. Faced with the demand for greater capacity, better performance, and even more advanced digital communication services, as well as competition from the allocation of new spectrum designated specifically for digital PCS, carriers were forced to implement the next generation digital standards. In the cellular band, this meant either a transition to cdmaOne, NA-TDMA, or temporarily to NAMPS.

Carriers around the world have enormous amounts of capital invested in AMPS infrastructure, and AMPS mobile stations can be developed and manufactured very inexpensively. This ensures that the AMPS system, and several other similar analog systems in use around the world, will probably stay with us for a while in the form of dual-mode standards and less and less in any type of standalone, exclusively AMPS network. Still today, there are many AMPS-only networks, particularly in rural areas, that have yet to implement, or for that matter even decide on, a digital transition.

3.15 Conclusion

The AMPS systems is one of the last major analog systems in use in the world, and it certainly is headed toward becoming obsolete some time in the future, considering the advent of digital formats. Still, AMPS has a tremendous footprint in the Americas, and in rural areas it is still often the only choice for mobile service. In addition, the tremendous investment carriers have made into AMPS infrastructure, which works adequately and brings in revenue, as well as government mandation for operators to maintain AMPS support, means AMPS will remain with us for a few more years.

Because AMPS networks generally also are the oldest systems, it should stand to reason that they generally require the most maintenance. In addition, AMPS is incorporated into both the CDMA and NA-TDMA standard. For these reasons, a very good understanding of the technology can help a technician greatly.

CHAPTER 4

NA-TDMA

4.1 Introduction to NA-TDMA

If you have read Chapter 3, you should understand that the AMPS system in the Americas had some shortcomings that had to be addressed. For instance, security was a serious problem, with an epidemic of cellular phone fraud costing carriers and users millions. Capacity was also an issue in high-use urban areas. Frequency channels were getting used up, and the FCC stated that it would not add any new spectrum for use by cellular companies. In addition, carriers wanted to give their customers similar advanced features in their mobile phones that they were getting with their landline phones, like call waiting, caller ID, and voice mail.

In Europe, GSM had been approved and deployment started. GSM solved many of the issues the AMPS networks had raised, such as fraud protection. GSM also had plans for advanced services, such as caller ID, short messaging, and, in the future, high data rates. In the Americas, however, carriers had invested billions in infrastructure and phones, and GSM was a system designed for a newly cleared spectrum band as a complete replacement. GSM was designed primarily for ease of roaming, and it was designed from a clean slate. American engineers felt that a new system needed to fit seamlessly into the already-present AMPS networks and architecture and give significant capacity increase.

While the Americas decided to come up with their own format, they did borrow much from GSM, including the very basic concept of using TDMA. Thus, the NA-TDMA system was born, incorporating the original AMPS with a new TDMA system, where TDMA channels would fit into AMPS channels. This book will refer to NA-TDMA, although it can also be called *Digital Advanced Mobile Phone Service* (D-AMPS), *American Digital Cellular* (ADC) and *North American Digital Cellular* (NADC), and very often it is referred to by the standard that defines it. Originally, the standard created defining NA-TDMA was IS-54. Later, some changes were added, including a digital control channel, support for the PCS bands, and some improved services. This new standard was labeled IS-136. This chapter will cover IS-136, as it is essentially an enhanced version of IS-54.

Thus, the new NA-TDMA system would encompass the following new features over AMPS, beginning with IS-54 and on to IS-136:

- Digital voice transmission;
- DMA and FDMA;
- All of the original AMPS protocols, including the same frequency channels (30 kHz);
- Increased capacity (theoretically three times more);
- Authentication for security;
- Better power consumption measures such as the sleep mode, which allows for longer battery life as well as less need for large cumbersome battery packs;
- Improved handoff systems (the MAHO system, where the handset helps make decisions on handoffs);
- Dual-mode mobile stations (AMPS and TDMA);
- Enhanced special services (such as caller ID and short message services);
- Support for the PCS band;
- Support for data transmission;
- A digital control channel.

4.2 Nomenclature of NA-TDMA—Identifiers and Architecture

Like all of the formats, NA-TDMA has its own terms for certain parts of its system. Because it is designed to be an evolution of AMPS, it uses the same identifiers and adds several new identifying codes. For instance, while NA-TDMA handsets have an ESN and MIN, they also have and *international mobile subscriber identification* (IMSI), which is basically the MIN but in a format that matches international conventions.

The architecture of NA-TDMA, while it can conform to IS-41, contains many proprietary features, just as AMPS networks do. The base station, MSC, and interworking function are therefore often grouped together and called the BMI, as the manufacturers can essentially set these functions up as they wish (see Figure 4.1). The interworking function refers to the interfacing between the base stations and MSC as well as some of the databases containing subscriber information.

Base stations have a couple of other identifiers they did not in AMPS. In addition to the DCC, a *digital verification color code* (DVCC) is added. This DVCC works the same way the SAT did in AMPS. Base stations also transmit a location area identifier. A specific LOCAID is assigned to a section of base stations, which then transmit it to the mobile stations. When a mobile station wanders into a new area, which it realizes because it sees a new LOCAID, the mobile station can then perform a registration. This allows the network to know a general area of where the mobile station is, so that it only needs to page a specific geographic area (or cluster of cells) if a call comes in for that user.

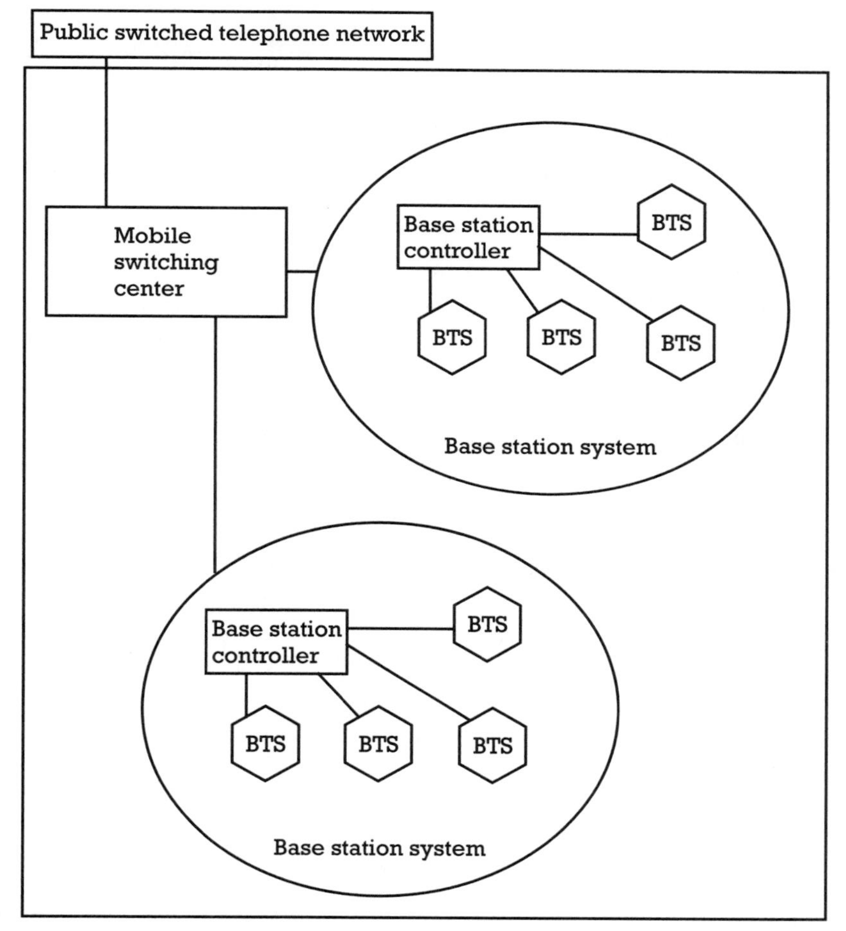

Figure 4.1 The base station, the MSC, and interworking function.

Another new identifier is the *system operator code* (SOC), which identifies the carrier operating a system. The SOC is actually transmitted by the base stations, but mobile stations can also be programmed with the SOC of their home station for roaming purposes.

Mobile stations also have several new identifiers. In addition to the station class mark, a *protocol version* (PV) has been added, also to relay capabilities of a specific mobile station. Also, as part of the enhanced security features of NA-TDMA, mobile stations have a unique 64-bit *authentication key* (A key) assigned to them by the carrier. This A key is never transmitted, and security of the network against fraud is dependent upon security of this A key. This will be discussed in detail later in the chapter.

4.3 Channels in NA-TDMA

NA-TDMA was designed to fit into the preexisting AMPS network easily. Thus, the digital channels have the same occupied bandwidth as an AMPS channel, 30 kHz. The control channels, whether analog or digital, also remain 30 kHz in NA-TDMA. IS-136 allows for AMPS FOCCs (for backward compatibility) as well as *digital control channels* (DCCHs). While the AMPS FOCC had a specific range of frequency channels to locate in, a DCCH can be located anywhere in the band.

Just as the name implies, transmissions are in a time domain in addition to frequency domain. There are two parameters needed to find the actual location of data—the frequency channel as well as the time channel (or slot). You will notice from Figure 4.2 that each 30-kHz channel on both the forward and reverse channel is divided into 40-ms frames containing six 6.67-ms time slots. During a call, the time slots on both the forward and reverse path correspond to each other (e.g., time slots 1 and 4 on the reverse path are the corresponding voice to time slots 1 and 4 on the forward path). The time slots are therefore offset from each other, so that time slot one on the forward path will begin 1.9 ms after time slot 1 on the reverse path is complete. In this way, even though communication is full duplex, a handset does not have to transmit and receive voice at the same time.

In the current deployment of NA-TDMA, full-rate channels are used. This means that two time slots correspond to one physical channel. In other words, a conversation might occupy frequency channel 285 and time slots 1 and 4. Pairs include one and four, two and five, and three and six. Thus, three conversations are possible

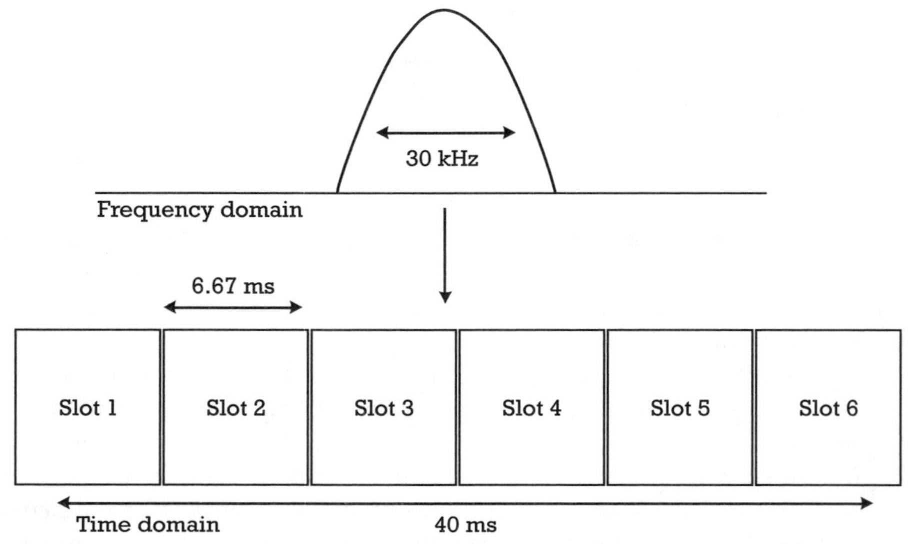

Figure 4.2 Each 30-kHz channel is divided into 40-ms frames, then subdivided into six 6.67-ms slots.

per 30-kHz frequency channel. NA-TDMA does specify a half-rate structure, which would allow one time slot per conversation, or six calls per 30-kHz channel.

If each conversation could use fewer time slots during each frame, there could be more conversations per frequency carrier. For instance, in full rate, three conversations would require six time slots. If a vocoder that only required one time slot per frame could be developed, this would double capacity. This is the goal of the half-rate systems (see Figure 4.3). Naturally, if less time is allocated to each conversation, the quality of the conversation might go down, unless an advanced vocoder was developed to compensate for this. The ability to move to half rate depends on development of a vocoder that will allow for similar-quality voice at lower data rates.

As you can guess, NA-TDMA also uses the same duplex spacing as AMPS. The forward and reverse channels are separated by 45 MHz in the cellular band and 80 MHz in the reverse channel.

Each time slot contains 324 bits of information. Each frame (six slots) therefore contains 1,944 bits. Because these 1,944 bits of information are transmitted every 40 ms, the data rate for each carrier is 48.6 Kbps.

4.4 π/4 DQPSK—The Modulation of NA-TDMA

The π/4 DQPSK is the modulation format used to get the 48.6 Kbps of information onto the carrier for transmission. Eight sine waves, each with a different phase, can

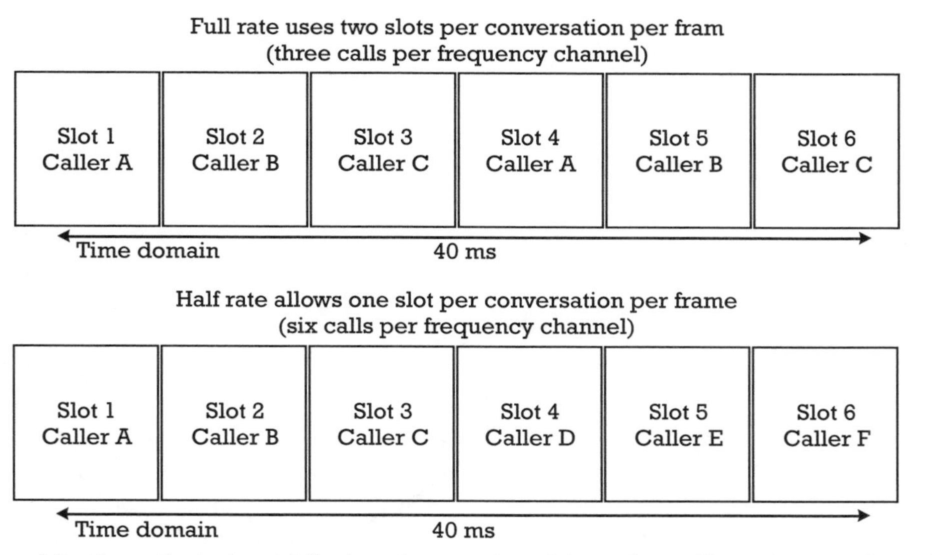

Figure 4.3 Currently deployed full-rate systems use two slots per frame (three users per channel), while a half-rate system would use one slot per frame (six users per channel).

be transmitted. These eight phase positions can be obtained with four phase changes:

- $\pi/4$ or 45°;
- $-\pi/4$ or −45°;
- $3\pi/4$ or 135°;
- $-3\pi/4$ or −135°.

The *differential* means that the system looks at phase changes and not the absolute phase position. This makes it easier for the demodulator, as it is easier to figure out a phase change than an absolute phase state (see Figure 4.4).

Each phase change refers to a channel symbol. With four possible channel symbols (the four phase angles just listed), each symbol can thus represent 2 bits each. Thus:

- $\pi/4$ or 45° = 00;
- $-\pi/4$ or −45° = 10;
- $3\pi/4$ or 135° = 01;
- $-3\pi/4$ or −135° = 11.

You can see from these four symbols that all sequences of ones and zeros can be generated. Thus, while the data rate for NA-TDMA is 48.6 Kbps, the symbol rate for NA-TDMA is 24.4 ksymbols/sec.

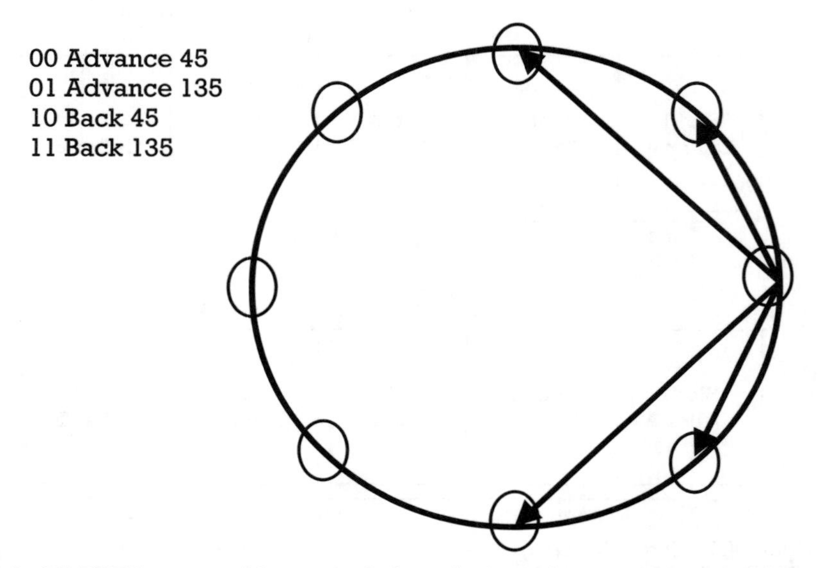

Figure 4.4 NA-TDMA uses two bits per symbol—each phase change represents a 2-bit pattern.

Another term you may have heard is modulation efficiency. Modulation efficiency refers to how many bits per second a system transmits in 1 Hz of bandwidth. NA-TDMA's modulation efficiency computes as such:

$$48.6 \text{ Kbps (data rate)}/30 \text{ kHz (bandwidth)} = 1.62 \text{ bps/Hz} \quad (4.1)$$

This is the highest modulation efficiency of all of the major systems in use today.

A higher modulation efficiency means less bandwidth is used to transmit the same amount of information (remember, a NA-TDMA digital channel needed to fit in 30-kHz channels—less bandwidth was important). One of the main reasons NA-TDMA has such a high modulation efficiency is that it does not use what is called *constant signal envelope* when making phase transitions. If you look at the constellation diagram of NA-TDMA, you can see that as it shifts from one phase state to another, it takes a direct path, causing the signal energy (the distance from the origin, or center of the constellation diagram) to change. If it used a constant signal envelope, the transition would move around the circle. The trade-off for this is that mobile stations will consume more battery power than other formats. GSM and FSK in AMPS do use constant signal envelope modulation and thus have lower modulation efficiencies. Thus, if you looked at the constellation diagram of GSM, what you would see would be a circle (phase changing, but amplitude remaining the same) (see Figure 4.5).

4.5 The FOCC, RECC, FVC, and RVC in IS-136

In the initial deployment of NA-TDMA defined by the standard IS-54, the control channels remained the same as they were in AMPS. The difference was that voice channels that supported TDMA were added (see Figure 4.6). With the need for PCS-band support as well as improved services rising, a DCCH was also introduced. However, the FOCC and RECC remain to allow dual-mode compatibility. This new standard became IS-136.

While the format for the FOCC, RECC, FVC, and RVC remains the same from AMPS to NA-TDMA, several additional messages support digital access, some improved services, and the other big benefit of the new system—authentication.

Four new messages on both the FOCC and RECC pertain to authentication. This is the process used to fight fraud and is discussed in detail later in this chapter. Note that authentication is not exclusive to NA-TDMA, but cdmaOne uses the process as well.

Two new FOCC messages, PAGE WITH SERVICE and MESSAGE WAITING, coupled with two new RECC messages, PAGE RESPONSE WITH SERVICE and ORIGINATION WITH SERVICE, are designed to allow a difference between voice

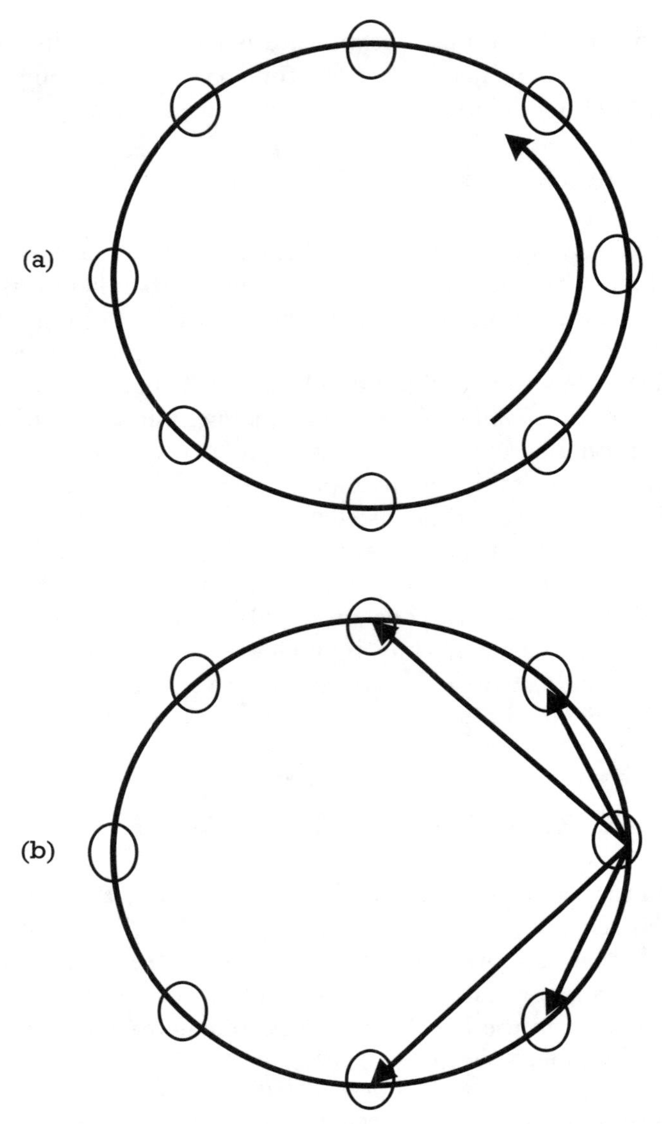

Figure 4.5 Constellation analysis shows the signal envelope of a system. Remember, distance from the center (or origin) is the amplitude of the signal. (a) A system that uses constant signal envelope would have shifts in phase, but amplitude remains the same. (b) NA-TDMA shifts the phase via a direct route, varying the amplitude in the process as shifts occur.

calls and data calls, including such features as short messaging and voicemail indications.

The final message added to the FOCC is probably the most important. This is the INITIAL DIGITAL TRAFFIC CHANNEL message, which, obviously, directs

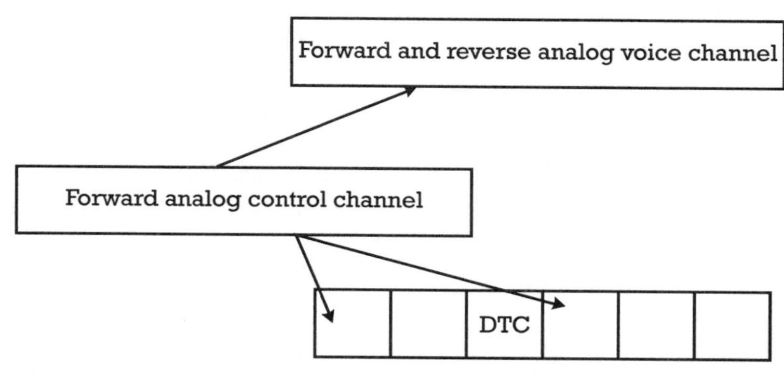

Figure 4.6 The original deployment of NA-TDMA, IS-54, used the same AMPS FOCC, but added a digital traffic channel.

the mobile station to a digital channel for a call. This allows systems to use the FOCC and divide up traffic channels between digital and analog. It was the format of the original deployment (IS-54).

The GLOBAL ACTION MESSAGE was part of the AMPS protocol, but in NA-TDMA it contains some new information. The most important piece of information in the GLOBAL ACTION MESSAGE is a DCCH pointer. This directs mobile stations to move to a DCCH for service, thus allowing for the advanced features of TDMA digital service (see Figure 4.7).

4.6 The Digital Traffic Channel

As discussed earlier, the *digital traffic channel* (DTCH) is divided into 40-ms frames. These frames consist of 1,944 bits, which are transmitted at 48.6 Kbps. Each of these frames is divided into six 6.67-ms slots, and two slots are used for each user (see Figure 4.8).

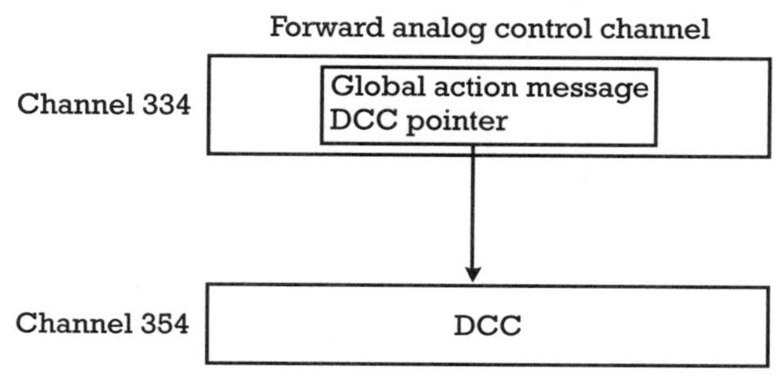

Figure 4.7 The GLOBAL ACTION MESSAGE contains a DCC pointer that helps the mobile stations find DCCHs (and obtain true TDMA digital service).

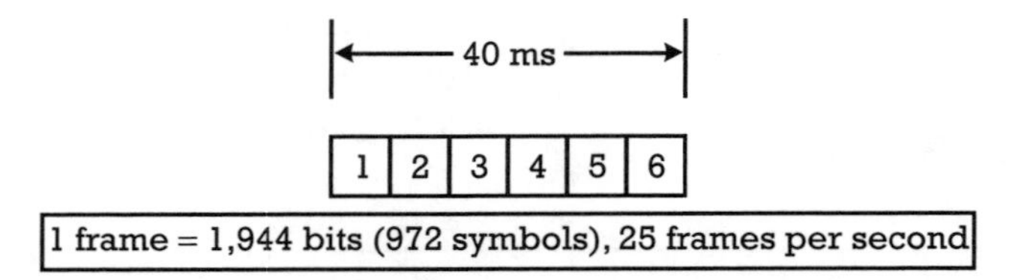

Figure 4.8 A digital traffic frame consists of six time slots, each with 324 bits, or 162 symbols (remember, two symbols per bit using $\pi/4$ DQPSK).

Built into each slot are a number of fields or *logical channels*. Figure 4.9 shows these logical channels for the forward and reverse DTCH. Keep in mind that the base station transmits the forward DTCH continuously, while the mobile station bursts its power on during its allotted time slot. It is also important for each mobile station sharing a frame to burst exactly in its assigned time slot. If mobiles are located at different distances from the base station, however, this could cause problems, as the propagation delay could cause an overlap into another time slot. For these reasons, the reverse DTCH places guard and ramp up bits at the beginning of the slot, while the forward channel simply begins with a synchronization stream.

Keep in mind that the information on the digital channels is protected against errors using convolutional encoding and the process of interleaving (see Figure 4.10). Convolutional encoding is a process using complex algorithms to provide a degree of error correcting. Interleaving is simply a method of scrambling bits so that when descrambled, if there were a series of lost bits in transmission, they would be spread out and not in a row.

The SYNC field naturally helps improve synchronization. It also serves a purpose in a process called *adaptive equalization*. While the specifics of adaptive equalization are well beyond the scope of this text, suffice it to say that it allows the receiver to compare the received waveform to what it knows it should be (the SYNC field will be known at the receiver) and compensate for differences.

Part of the synchronization process involves setting a parameter called *time alignment*. As mentioned earlier, the mobiles are spread out at different distances from the base station, causing the signals to take different amounts of time to get to

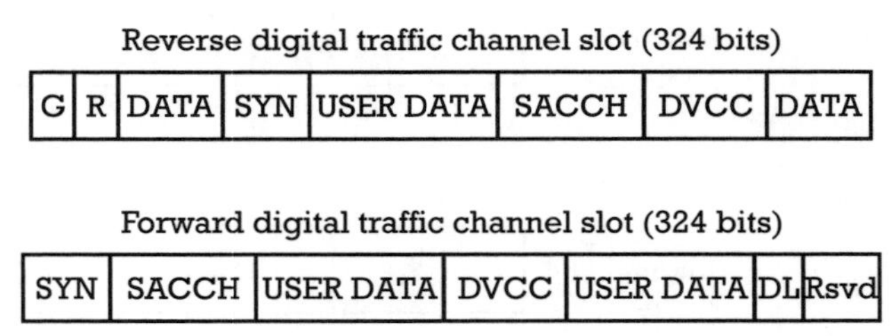

Figure 4.9 The individual slots contain various fields (besides the user's data) called logical channels.

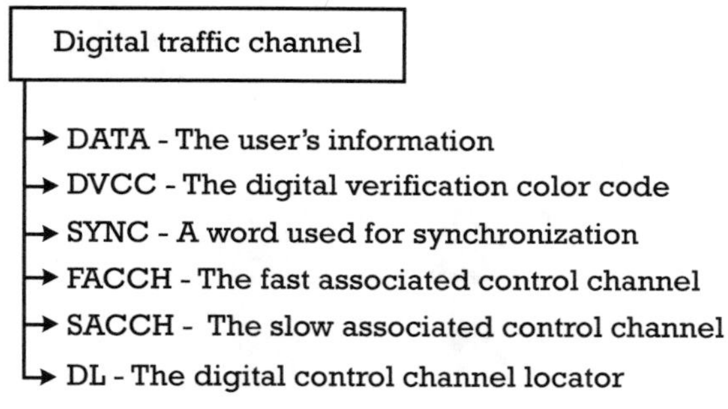

Figure 4.10 The various logical channels that can be placed in a digital traffic channel.

the antenna. Time alignment allows for the mobile to turn its transmitter on at a specific offset so that it hits the base station's antenna at the right time, not causing any interference to a neighboring slot.

To assist in this process, it is possible for a mobile station to transmit a shortened burst. The shortened burst uses a guard bit field of 50 bits, as opposed to the normal six. This shortened burst will then fit in its assigned slot, even if the time alignment is out significantly (see Figure 4.11). The shortened burst transmits only the SYNC and DVCC (with zeros filling the rest of the slot) repeatedly, until the base station can calculate the correct time alignment value.

Two other logical channels on the DTCH are the DVCC and the *slow associated control channel* (SACCH). The DVCC is very similar to the SAT, in that the base station transmits it and the mobile station must receive and retransmit back to the base station. The SACCH allows messages to be sent during the conversation, without preempting any of the data. You might remember from the AMPS chapter that any signaling that took place was in the blank and burst mode, in which the voice was preempted and an FSK data stream was transmitted whenever a message was sent.

The messages that need to be sent on the SACCH are 132 bits long. Because the SACCH is only 12 bits per slot, that means it takes 11 slots, or six frames, to get the entire message through. Six frames would equal 240 ms. Sometimes, this is not fast enough because certain operations (e.g., handoff operations) require a much quicker response. For this purpose, another associated logical channel, the *fast associated control channel* (FACCH), is introduced. The FACCH replaces the DATA field and is thus 260 bits per slot. Like the blank and burst in AMPS, the FACCH does preempt voice, but it performs a trick called *bad frame masking*, which basically causes the receiver to repeat the last received blocks, thus limiting the annoyance of the FACCH.

The DATA field in the DTCH is the vocoded voice. NA-TDMA now uses the ACELP vocoder. This vocoder divides the voice information into 20-ms blocks, and each 20-ms block is represented by 260 bits. If you divide 260 bits per speech block

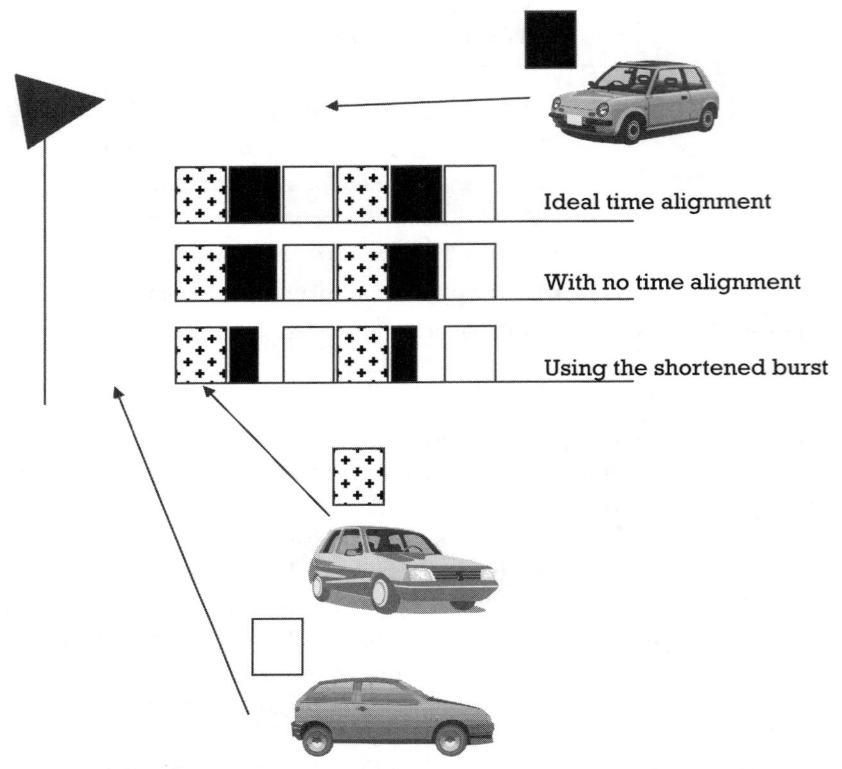

Figure 4.11 Time alignment is an important element of quality service in any TDMA system. The shortened burst allows a method of transmitting while time alignment is computed without interfering with other channels.

by 20 ms per speech block, what you will get is 13 Kbps, the transmitted speech bit rate of the ACELP vocoder.

The final logical channel on the forward DTCH is the *digital control channel locator* (DL). The DL basically tells the mobile station the channel number of a DCCH, which the mobile station can use after the conversation is over.

4.7 Messages on the DTCH

Messages on the DTCH arrive and depart on either the SACCH or the FACCH. For the most part, these messages are very similar to messages already discussed in the AMPS chapter (e.g., the RELEASE and ALERT WITH INFO messages). The ALERT WITH INFO becomes increasingly important as this is the message used to deliver such PCS services as caller ID. In addition, many carriers have deployed text messaging services, and messages called R-DATA are used to send and receive these messages.

PCS services often require the user to interface with the switch by pressing the various numbers on the mobile station. For instance, many voicemail systems require the *dual-tone multiple frequency* (DTMF) tones to listen, save, or erase messages. While these tones are easily transmitted in analog transmission, NA-TDMA puts its audio through a vocoder. Because a vocoder would distort these tones during the encoding/decoding process, a different system was needed to transmit DTMF tones. Thus, SEND DTMF messages are used on the associated channels.

You will remember that the base station needs to closely control the timing of the mobile station in addition to the power levels. This control is performed by using a PHYSICAL LAYER CONTROL message.

A variety of messages is associated with the authentication process and MAHOs, which will be discussed in detail later in this chapter.

4.8 The DCCH

The format of the DCCHs is a bit confusing, to be sure. First, it is organized into 1.28-second hyperframes. These hyperframes contain two 0.640-second superframes. Within these superframes, there are sixteen 40-ms frames. Each frame consists of two 20-ms blocks, and each block contains three 6.67-ms slots. Like the DVCH, each DCCH will use two slots per frame in full rate, and one slot per frame in half-rate. Confused yet? Figure 4.12 may make it a bit more clear.

The structure of the slots is very similar to the DTCH, except that the reverse DCCH requires more synchronization than the reverse DTCH, as transmissions are not as often or patterned as they are during a conversation. Thus, the base stations needs a few extra "hints" in order to demodulate the bursts correctly (see Figure 4.12).

Like the voice channels, the DCCH has a variety of fields and logical channels multiplexed on it (see Figure 4.13). The SYNC is again used for ensuring correct timing, with the *preamble* (PREAM) adding the extra hint. The *superframe phase* (SFP) is placed in the forward DCCH and gives the mobile station the current block number within the superframe. This is useful not only from a timing standpoint, but also because this field is in the exact location as the DVCC in the forward DVCH, so a mobile can examine this field and quickly determine if it is a DCCH or a DVCH.

The *shared channel feedback* (SCF) field actually contains three subfields used by the mobile to help with reverse channel transmissions. The *busy/reserved/idle* (BRI) field is similar to the busy/idle bits in AMPS. It simply informs the mobile station if the reverse slot, called a *random access channel* (RACH), is being used. A mobile station will monitor this field; once it goes idle, the mobile station can transmit its reverse DCCH in a specified RACH. Three frames later, two other fields will provide confirmation to the mobile station of a successful transmission. The *received/not received* (R/N) field will indicate a result, as well as the *coded partial*

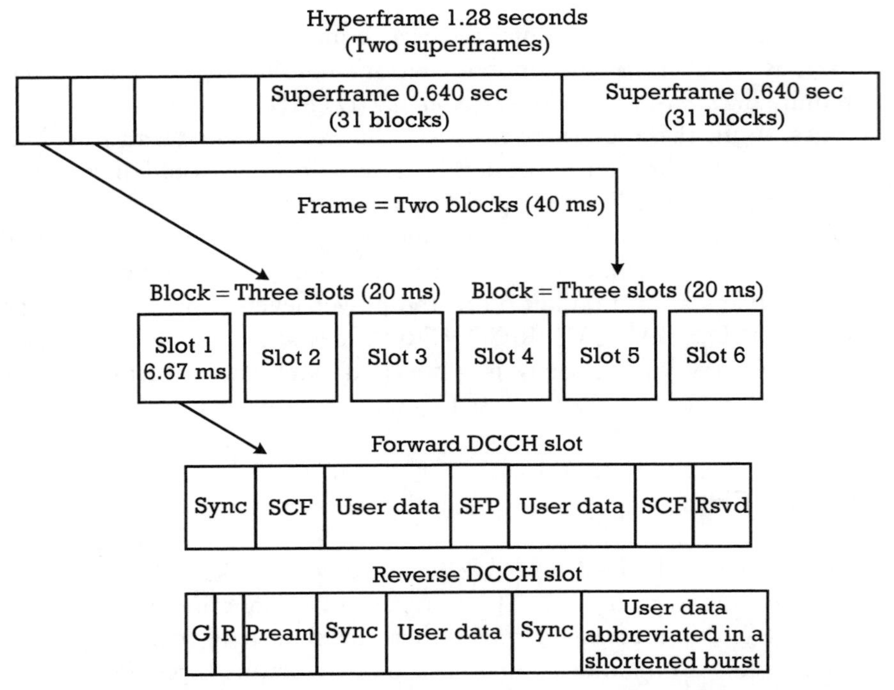

Figure 4.12 The DCCH uses an elaborate hierarchical structure that allows for different multiplexing options as well as the NA-TDMA handset's sleep mode.

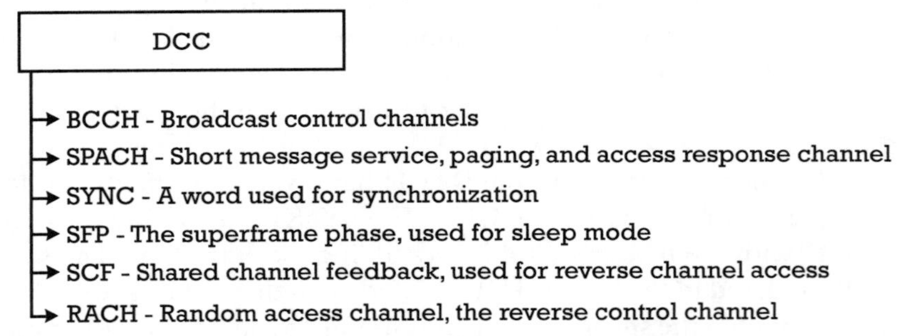

Figure 4.13 The DCCH has various logical channels carrying the various information that handsets and networks need from each other.

echo (CPE) field. The CPE is actually a part of the MIN of the transmitting mobile station, so the mobile station can be sure it was the mobile station to which the base station was listening.

The data on the forward DCCH can be one of two types of channels: a *broadcast channel* (BCCH) or a *short message, paging, and access response channel* (SPACH). Close to 60 different messages can be transmitted on the DCCH, as opposed to the 16 in AMPS. This is due to the enhanced services and controls present in IS-136 (e.g., authentication alone adds at least nine messages).

Broadcast messages are naturally intended for all of the mobiles in the cell. There are three types of broadcast messages:

- *Fast broadcast control channel* (F-BCCH);
- *Extended broadcast control channel* (E-BCCH);
- *Short message service broadcast control channel* (S-BCCH).

4.9 Sleep Mode

The reason for the hierarchical structure of the DCCH is to facilitate the sleep mode. If the mobile was to monitor the control channel continuously, as it does in AMPS, the standby time of the mobile station would be lower than if the mobile station only had to monitor the control channel periodically. This sleep mode is how digital handsets can dramatically increase the standby time over their analog ancestors.

Calls coming in for the mobile station will arrive in paging subchannels in the SPACH blocks and are transmitted twice—one in each superframe (remember there are two superframes in each hyperframe). Essentially, sleep mode works by allowing the mobile station to wake up in time to listen to specific SPACH messages in specific hyperframes. Assigning which paging channel a specific mobile station will listen to is a function of the MIN of the mobile station, and this allows for a pseudorandomness in the distribution of the paging channels. How much the mobile stations "sleep" is a function of a parameter called the *paging frame number* (PFN) (see Figure 4.14). The paging channel for a specific mobile station will occur in a specific hyperframe. If the PFN equals one, that means there is an assigned paging channel

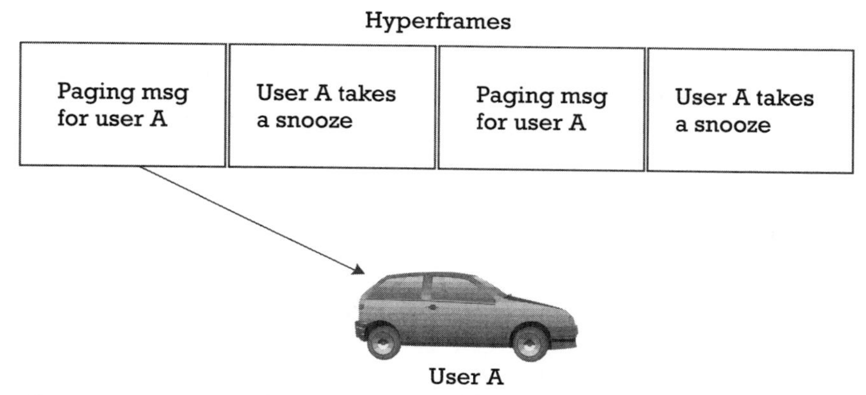

Figure 4.14 During sleep mode, the mobile station can take a nap, waking up in time to read its assigned hyperframe. Here a phone is assigned PFN of two, waking up every other hyperframe.

in each hyperframe. As each hyperframe is 1.28 seconds, that would mean the mobile station could only sleep in between each hyperframe, or 1.28 seconds. The highest PFN, 96, means the mobile station can sleep through 96 hyperframes before it sees the hyperframe assigned to it, a very long 122.88 seconds, which would give a very long battery life but lousy service, as a caller would have to wait through two minutes of ringing before the lazy phone received the paging message.

4.10 Messages on the DCCH

As discussed earlier, three logical channels are used to transmit control channel messages. On the forward DCCH, there are the BCCHs, as well as short message service, paging, and access response channel. The third logical message channel is for the reverse DCCH, the RACH.

It was also mentioned earlier that there are about 60 messages on the DCCH. Naturally, many of the messages are similar in function to those on the FOCC discussed already in AMPS. The majority of the broadcast messages are intended to provide the overhead information and configuration parameters that all mobile stations in the cell will need. Some of the common BCCH messages include:

- SYSTEM IDENTITY: This identifies the base station, including the SID, PV, and network type.
- DCCH STRUCTURE: This contains information on how the superframes are organized, as well as the DVCC.
- CONTROL CHANNEL SELECTION PARAMETERS: This gives the mobile station information about other DCCHs in the cell.
- ACCESS PARAMETERS: Here, the base station gives the mobile stations some of the variables needed to complete reverse channel access on the RACH (e.g., the number of attempts to gain access before giving up).
- SOC/BSMC IDENTIFICATION: This identifies the carrier as well as the manufacturer of the base station.
- EMERGENCY INFORMATION BROADCAST: This is a broadcast text message to all of the users (e.g., "Go Cowboys!" ...well, perhaps for more important matters than football teams).
- REGISTRATION PARAMETERS: This sets up registration procedures in the cell.
- TIME AND DATE: What else? The time and date!

SPACH messages are the messages aimed at specific mobile stations. Here we find the PAGE, RELEASE, MESSAGE WAITING, CAPABILITY REQUEST, registration responses (either accept or reject), DIGITAL VOICE CHANNEL DESIGNATION,

ANALOG VOICE CHANNEL DESIGNATION (mobile stations can use a DCCH, but get assigned to an analog voice channel), and DIRECTED RETRY messages, which are all either similar to AMPS messages or self-explanatory.

Just as in AMPS, the reverse control channel only has a few messages. Like AMPS, these include ORIGINATION, PAGE RESPONSE, and REGISTRATION messages. A response to the previously mentioned SPACH message CAPABILITY REQUEST, the CAPABILITY REPORT tells the base station its capabilities. There are several other messages that again relate to authentication and MAHO, which will be discussed separately.

4.11 MAHO

Free time is something we are all after, and NA-TDMA gave mobile stations some serious free time. If you remember, there are six time slots in each 40-ms frame, but a mobile station only uses two of those time slots during any given job. This leaves four time slots, or 26.67 ms, for the mobile station to relax and enjoy itself. Unfortunately, the designers of NA-TDMA had other ideas and decided to put the mobile station to work during these down times. This leads to the process of MAHO.

In AMPS, the base station monitored the signal quality of a call and made determinations on handoffs as such. MAHO allows for the mobile station to monitor signal quality as well and thus can help make better decisions on when and where to make a handoff. MAHO also allows the mobile station to monitor other channels on different base station and pick one that would work best. This makes the handoff process much more efficient and effective. Also, poor quality on a channel is often not due to signal strength, but rather interference. In MAHO, the digital method of measuring reception, *bit-error rate* (BER), can be measured by the mobile station. BER is a good indicator of interference and is much more indicative than a signal strength measurement alone.

During a call, the mobile station measures the BER and signal strength on the surrounding base stations while it is not transmitting or receiving in its slot. It reports this information in the CHANNEL QUALITY message (using the SACCH). The mobile station is told where to look for channels via a message called MEASUREMENT ORDER, which includes 6 or 12 different frequency channels it should measure. The CHANNEL QUALITY message then sent from the mobile station includes the channel measured, the BER, and the RSSI (see Figure 4.15).

Part of MAHO is a procedure called *mobile assisted channel allocation* (MACA). Before a call is set up by the control channels, the base station broadcasts a MACA message to all of the mobile stations in the cell. This MACA message contains all of the idle channels in the cell that could use a call on them. During off time in the idle mode, the mobile station then takes a look at these channels and let the network know the results via the MACA REPORT on the reverse DCCH.

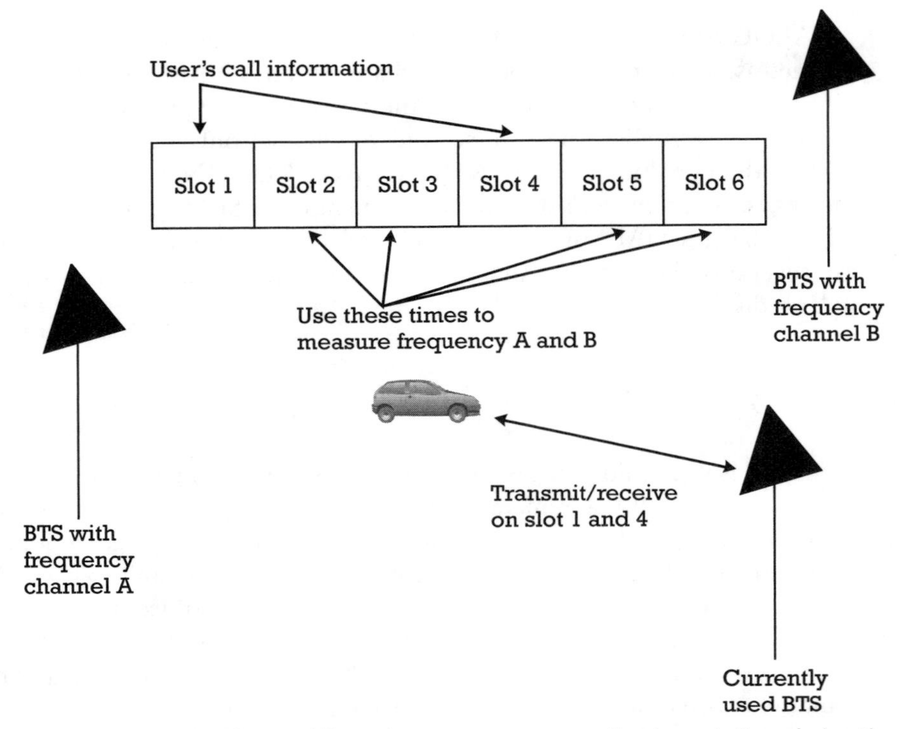

Figure 4.15 During MAHO, a mobile station measures surrounding base stations during time slots it is not using.

4.12 Authentication

One of the biggest concerns of today's cellular operators is the massive fraud that has been robbing the carriers and cellular users of an enormous amount of money. While precautions were taken to provide some measure of protection in AMPS, the ingenuity and determination of today's high-tech criminal overcame most of these systems to the point that cellular fraud-enabling technology is nearly commonplace. While no system is foolproof, new systems can be developed to make fraud as difficult as possible. Authentication is just such a process—not foolproof, but certainly a major hurdle for the common criminal (perhaps closer to impossible for the "common" criminal).

The authentication is standardized in IS-41-C (the networking standard) and thus can be independent of an air interface; it is incorporated into both IS-95 and IS-136. Because NA-TDMA was the first air interface to incorporate support for this process, it is included in this chapter. Keep in mind the process is the same across air interfaces, though, and this text will reference back to this section in the CDMA chapter.

Authentication is a several-step process, and it occurs many times during a normal mobile stations day, most notably during registrations, call origination, and call termination whenever the mobile station moves to a new location. The process itself is very complex (on purpose) so this text will give a very basic overview of the five basic functions involved in the process (a sixth process, authentication reporting, is much more of a network management process and is thus left out):

- Shared secret data sharing;
- Global challenge;
- Unique challenge;
- Shared secret data update;
- Call history count update.

First, a review of the terms is needed:

Global challenge is an authentication procedure on all mobile stations attempting to access a particular base station. It is called the global challenge, as opposed to the unique challenge (discussed later), because the challenge and *random number needed* (RAND), are sent as broadcast messages on the control channel—thus to all of the mobile stations that are monitoring that particular control channel.

Unique challenge is an authentication procedure directed at a particular mobile station that may be trying to gain access, or perhaps is already on a call. The process is similar to the global challenge, except it is not performed in a broadcast manner.

The *authentication center* (AuC) is the database in the network architecture that controls much of the private authentication data and performs many of the complex calculations.

The A key is a 64-bit number stored permanently in each mobile station that supports authentication (many older phones do not). It is permanently assigned to the mobile station and securely stored in the mobile station and in the authentication center, such that it cannot be modified nor even known by the user. The secrecy of the A key is really the key to the entire process being a successful blockade against fraud. It is never transmitted over the air nor transmitted between network functions (i.e., from one system to another).

Shared secret data (SSD) is a 128-bit number actually composed of two separate 64-bit numbers: SSD_A and SSD_B. SSD_A is used for authentication purposes, while SSD_B is actually for encryption of voice or specific signaling messages. The SSD is generated using the A key as well as the ESN of the mobile station and a random number the network generates. This allows the SSD to be changed from time to time (particularly if the network operator feels someone has compromised system security) via the process called SSD update.

SSD update is an authentication process that involves changing the SSD in the AC as well as the mobiles. Part of this process is the sharing of a new random number used to generate the SSD (called *RANDSSD*), another process called base station challenge (to make sure the base station requesting this change is real), and a unique challenge to make sure the SSD update was correct.

Call history count update, the COUNT value, is a parameter stored in the mobile station as well as the network AC, which provides an additional layer of protection against fraud. Basically, the network controls this parameter through an update message, and the value in the network should match that in the mobile (or at least be close—a certain degree of error is understood due to transmission errors and such). The system uses the PARAMETER UPDATE message on either the traffic or control channel to increment the COUNT value. Thus, if criminals do manage to decipher the authentication process and clone a mobile station, they still would have to stay close to the COUNT value in the cloned mobile station's home system, a nearly impossible task.

CAVE algorithm is the complex algorithm (let us call it a black box) that takes an input and generates various new codes needed for the authentication and encryption process. Both the mobile station and AC have CAVE black boxes. The CAVE algorithm uses the A key, ESN, and a random number called RANDSSD to generate the SSD. It also uses the ESN, part of the MIN, SSD_A, and a random number (either RAND, RANDU, or RANDBS, depending on whether it is a global challenge, unique challenge, or base station challenge) to generate the authentication codes (AUTHR for a global challenge, AUTHU for a unique challenge, and AUTHBS for a base station challenge).

The global authentication challenge process involves several steps. First, the network generates a random number called RAND. This RAND is sent to the mobile station, which uses it, the ESN, the MIN, and the SSD_A via the CAVE algorithm to generate an authentication result code, called AUTHR, which is sent back to the network. The network performs the same computation, generating its own AUTHR, and compares the received to its own, making the decision on validity and allowing access (see Figure 4.16).

Thus, the access message from the mobile station includes AUTHR, the value of its COUNT register (defined earlier), and the eight most significant bits of the random number the mobile station received from the network (RAND) called RANDC. The reasons for transmitting RANDC is to ensure that the mobile station isn't simply a fraud replaying a message it recorded, that the RAND the mobile station is using is wrong because it was received from a neighboring system, or that the RAND was wrong because the mobile station simply didn't receive it correctly.

The unique challenge is very similar to the previous description, with some slight changes in the variables. As mentioned earlier, the unique challenge differs from the global challenge in that it is directed at a specific mobile station, whereas the global challenge is aimed at all mobile stations globally. The unique challenge often serves as the double check of the global challenge—a second opinion, if you will. Also, the unique challenge's random number, RANDU, is changed each time a unique

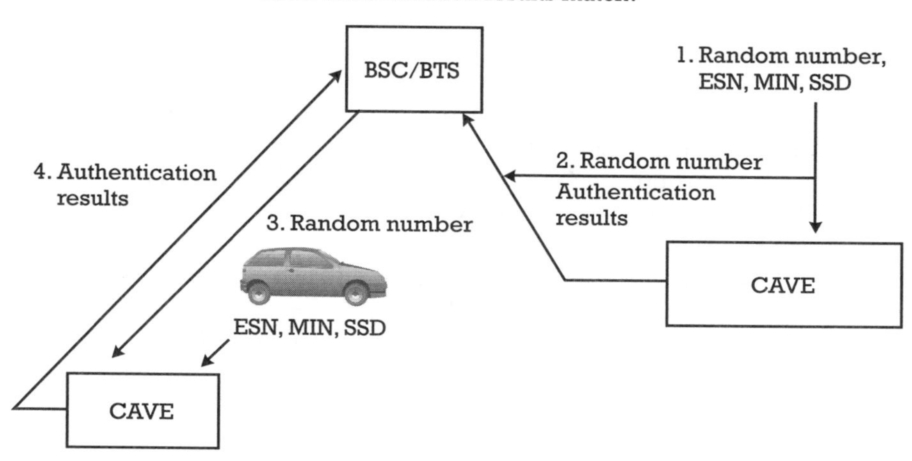

Figure 4.16 The authentication process, whether during a global or unique challenge, involves sharing a random number, using the CAVE algorithm to compute an authentication result, and comparing this result at a central location for access.

challenge is initiated, which makes it a substantially more secure challenge. Generally, a unique challenge will be ordered if the global challenge fails. The unique challenge is ordered after a global challenge as a double check in areas of very high fraud on specific call originations or terminations or simply on a timer basis.

The SSD update process is actually several combined processes. First, when a mobile station receives the order to update its SSD, it needs to make sure that the base station requesting this is for real, not fraudulently trying to get information from the mobile stations or simply trying to sabotage the system by changing mobile station's SSDs (the mobile stations would thus lose service unless the network discontinued authentication for a time). The mobile station does this via a base station challenge.

This is basically the same process as the unique challenge but backwards. In this case, the mobile station generates the random number, RANDBS, using it to generate the authentication code, AUTHBS. The network receives the RANDBS and generates its own AUTHBS, sending it back to the mobile station for comparison. Once the mobile station is sure there is a match, it can move on to updating its SSD.

Part of the original order to begin this process was yet another random number, RANDSSD, which was generated at the AC and sent to the mobile station. This RANDSSD is combined with the A key and the ESN of the mobile station, which then creates the new SSD. The network as well as the mobile station should have performed the same calculations and arrived at the same new SSD. To ensure this is the case, the process is not complete until a unique challenge is initiated and completed.

4.13 Conclusion

The NA-TDMA system was designed as an upgrade path for analog systems in North America. Designed to incorporate the AMPS standard, AMPS originally began using the AMPS control channel, then later added a true digital control channel with the implementation of IS-136. With many large carriers operating NA-TDMA systems, it certainly will remain an important system in the Americas. However, as many operators are looking for cost-effective ways of getting to the future by beginning to implement GSM on their NA-TDMA networks, it remains to be seen how long NA-TDMA will remain a force. Like AMPS in the 1990s, however, the sheer volume of NA-TDMA infrastructure and phones across the Americas ensures that it will remain important for the time being.

CHAPTER 5

GSM

5.1 Introduction

Groupe Special Mobile first started using the initials GSM to represent the new digital standard they were tasked with developing. Conveniently, the initials came to stand for the Global System for Mobile Communications. The phenomenal growth of the format throughout the world is quite unprecedented, as some form of GSM is used on every inhabited continent.

The goals of GSM were simple: to create a digital cellular system in which international roaming was seamless with a variety of enhanced services. In support of the project, the European community universally allocated new spectrum specifically for the all-digital system. Unlike NA-TDMA, GSM would not need to support the already-built analog network, but would be a complete replacement. This vision has been achieved for the most part, and GSM continues to grow throughout the world.

There are several varieties of GSM in use throughout the world. The primary difference is the frequency band being used. For instance, in the Americas, GSM is used in the PCS band and is often referred to as PCS-1900. GSM in Europe uses the all-digital 900-MHz band. In addition, an 1,800-MHz band has been allocated, and GSM in this band is often called DCS-1800.

In the end, the primary advantages of GSM over the earlier analog systems include:

- Support for international roaming;
- A distinction between user and device identification;
- Improved speech quality;
- Extensive security;
- Additional new services (e.g., short messaging and caller ID).

GSM is regulatedby ETSI, which first adopted the standard in 1991. Unlike other standards, GSM is a standard in the true sense of the word. The entire system is standardized, including all of the interfaces between the various components of the system. This is in sharp contrast to the North American systems, where most of the interfacing is proprietary.

5.2 GSM System Architecture

Most of the North American standards borrowed much of the terminology of their systems from GSM. The GSM refers to the standardized interfaces specifically as is shown by Figure 5.1. As with other formats, base stations are referred to as BTSs, which are connected to a *base station controller* (BSC), which manages the radio resources for all of the BTSs. The BSC and BTSs are referred to as the *base station subsystem* (BSS), and the BSSs are all linked to an MSC. This MSC serves as the gateway to both the outside world (e.g., the PSTN/ISDN) as well as to the various databases needed to manage users on the network, such as the HLR, the VLR, the AC, and the EIR.

The HLRs and VLRs aid the network in the primary design goal of GSM—international roaming. Here the information and location of various mobile stations is stored so that the network can route calls appropriately. The EIR is a database that stores the *international mobile equipment identity* (IMEI) of all the mobile stations

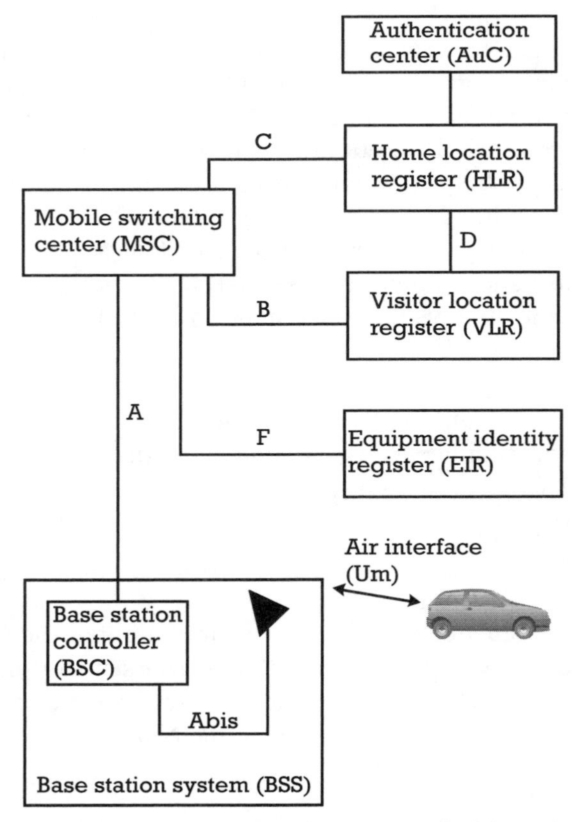

Figure 5.1 The GSM network architecture. Note that in GSM all of the interfaces are standardized (A, B, C, D, F, Abis, and Um interfaces are shown).

in the network. The network can look at this list and decide if a mobile is stolen or not very quickly. The *authentication center* (AuC) is used for authentication. GSM mobile stations use a *subscriber identity module* (SIM) card that contains a special key. A copy of this key is kept at the AuC for verification before calls are placed.

5.3 Identifiers in GSM

There are several different identifiers in GSM than are seen in CDMA or NA-TDMA. Many of these identifiers are stored on the SIM card, which means the end user can move to a different phone by simply removing the SIM out of the old phone and inserting it into the new one. Because the SIM card contains the unique identifiers, the user would still retain the same phone number, service plan, and perhaps even the directory database he or she has set up for speed dialing.

These identification codes now have three possible origins: the SIM card, the network, and the mobile station hardware itself (see Figure 5.2).

Here are some of the identification codes:

- *IMEI:* This 15-digit serial number is assigned to the mobile station at the time of manufacturing.
- *IMSI:* This is the phone number assigned by the carrier to the user. It is stored on the SIM. The 15-digit IMSI takes into account country codes to allow for international service.

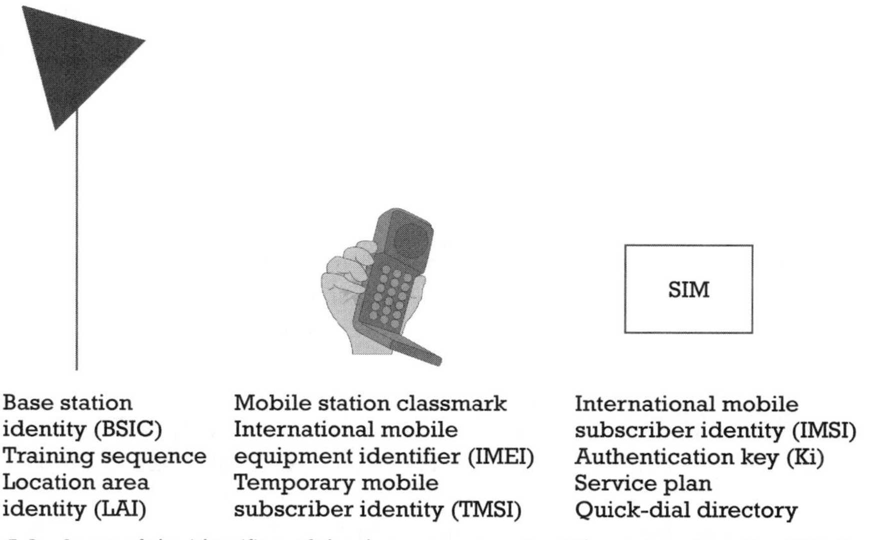

Figure 5.2 Some of the identifiers of the three components of the air interface: the BTS, the mobile station hardware, and the SIM card.

- *Temporary mobile subscriber identity (TMSI):* This identifier is assigned by the VLR after a mobile station establishes itself in the network. The network then uses this identifier rather than the IMSI when performing various call management tasks. This identifier is shorter than the IMSI (which makes it more efficient to transmit), and it adds a bit of security because the IMSI does not need to be continuously transmitted.
- *Authentication key (Ki) and cipher key (Kc):* The Ki is assigned to the end user and resides on the SIM. Similar to the authentication process in NA-TDMA and CDMA, the Ki is never transmitted but rather is used to compute a ciphering key (Kc). Kc is then used to protect the transmitted information and prevent any unauthorized interception of the information.
- *Mobile station classmark:* This classmark, like those in other systems, tells the network the capabilities of the mobile station. This includes the version of protocol it supports, the power levels it can support, its encryption capabilities, the frequencies it supports, as well as the mobile station's ability to support special services.
- *Location area identity (LAI):* This identifies the country and system of the base station. It is similar to the SID in the Americas.
- *Base station identity code (BSIC):* This code, in addition to a training sequence, is assigned to each base station. It serves a similar purpose as the SAT in AMPS, the DVCC in NA-TDMA, and the PN offset in CDMA—to identify the particular base station, ensuring the mobile station it is listening and transmitting to is the right base station. Like the SAT, the mobile station retransmits the code to the base station in order to close the loop.

5.4 Modulation in GSM

As described in Chapter 1, GSM uses a modulation format called GMSK (see Figure 5.3). The transmit rate of the GSM system is 270.833 Kbps, while the bandwidth of the signal is 200 kHz. Thus, the modulation efficiency of GSM (data rate divided by bandwidth) is 1.35 bps/Hz. This is a lower efficiency than NA-TDMA (1.6 bps/Hz). One of the trade-offs for the lower modulation efficiency is that GSM uses a constant signal envelope, which means less battery drain and more robustness in the presence of interfering signals.

In having a constant signal envelope, the constellation diagram of a GSM signal is a circle, and, thus, unlike NA-TDMA and CDMA, constellation analysis will not tell a technician very much about the quality of modulation.

Another important difference between GSM and NA-TDMA pertains to the downlink transmission. In NA-TDMA, the base station transmitted all slots

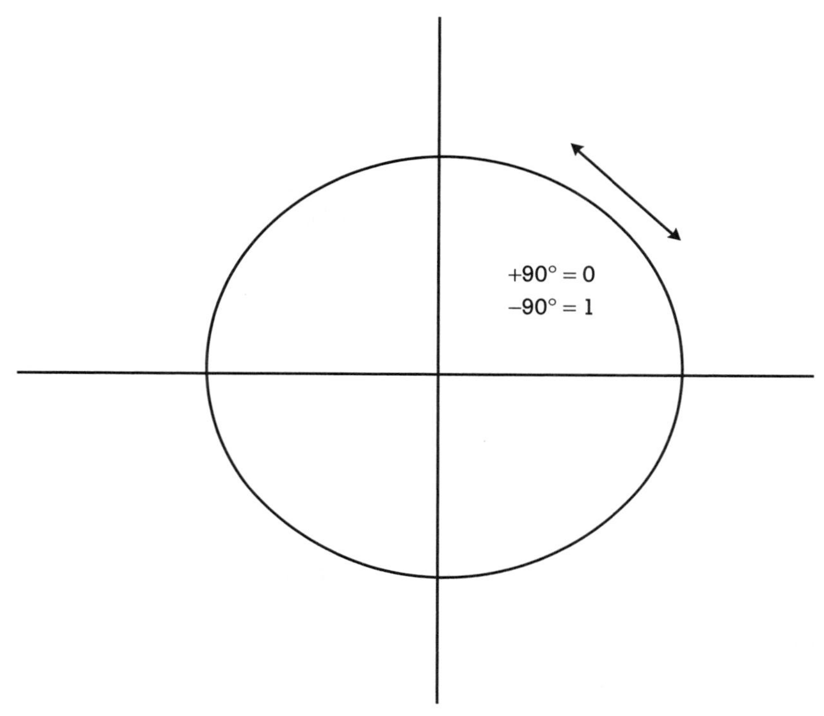

Figure 5.3 GSM uses GMSK modulation. Amplitude remains constant during phase shifts of ±90°. The constellation diagram of a GSM signal thus resembles a circle.

continuously, regardless of whether the slots were assigned or not. GSM specifies that the base station can turn the transmitter off if the slots is not needed.

5.5 Power Levels

Like the American standards, GSM can use a variety of power levels and can change these power levels during a call. The GSM standard allows for five classes of mobile stations divided by the maximum amount of power they can output (see Figure 5.4). The highest class outputs 20W (43 dBm), and the lowest class outputs 800 mW (29 dBm). Below the maximum power are 16 power levels with 2-dB steps (a dynamic range of 30 dB). Typically, mobile stations have a maximum power of 2W, but as terminals only transmit one eighth of the time (one burst of the eight), the average maximum transmit would be 250 mW.

GSM also uses a process called *discontinuous transmission* (DTX). It is important to limit the amount of power transmitted in the system—cochannel interference is a primary cause of suboptimal capacity—so limiting the overall output power is

Power control level	Power class*	Peak power (dBm)
2	1	39
3	2	37
4	3	35
5	4	33
6		31
7		29
8		27
9		25
10		23
11		21
12		19
13		17
14		15
15		13

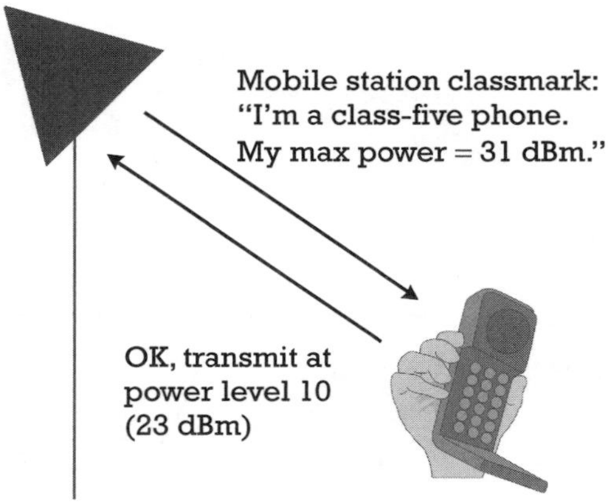

Mobile station classmark: "I'm a class-five phone. My max power = 31 dBm."

OK, transmit at power level 10 (23 dBm)

There are five power classes in GSM; these indicate the peak power in the class. All are capable of the levels below their peak ratings.

Figure 5.4 The various power classes and the 15 power levels used in GSM.

important. An equally important reason to limit transmit power is to save battery life. One way of doing this is to use the lowest power level needed to communicate without errors. DTX is another method of lowering power output. Similar to the variable vocoder used in CDMA, the DTX system detects if there is voice activity (of course, distinguishing between voice and noise is an important consideration). If there is no voice activity, then the transmitter can shut down. The other intrinsic benefit of DTX is that the talk time of the handset will increase dramatically. An interesting part of this process is the concept of comfort noise. If the transmitter is turned off at the mobile station, the other party would hear dead silence, a rather disturbing sensation. To help the situation, engineers came up with the concept of comfort noise, to ensure the other party that the call is still connected.

5.6 GSM Channels

Like all of the main cellular and PCS formats, GSM uses separate frequency bands for the uplink and downlink. The channels occupy 200 kHz. Like NA-TDMA, GSM divides a frame into slots. In this case, the frame is 4.62 ms long, and it is divided into eight time slots. There is an offset between the related slots on the uplink and downlink to allow the mobile station to transmit and receive at different times.

The base station will transmit two types of channels to the mobiles: traffic channels and control channels. The channel structure is organized into multiframes, with 26 frames making up a traffic channel multiframe and 51 frames making up a control channel multiframe (see Figure 5.5).

In the 26 frames that make up the traffic multiframe, 24 are used for voice traffic: frames 0 to 11 and 13 to 24. An SACCH (discussed later) is inserted in either frame 12 or frame 25. The remaining frame is left unused in full-rate speech operation (see Figure 5.6).

In half-rate speech operation, each frame can support twice as many callers (16); however, one SACCH can still only support eight slots per frame. This means that if every caller needs a SACCH associated with the conversation per multiframe, an additional frame is needed as a SACCH. Hence, the remaining frame comes into play and is used exactly for this purpose.

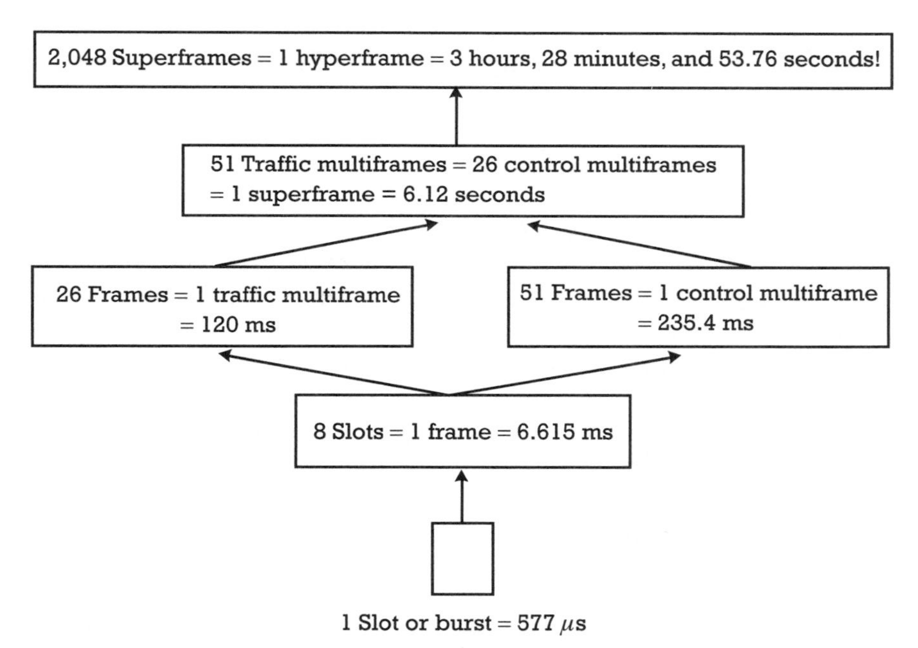

Figure 5.5 The GSM frame organization. This structure allows for various types of channels to be inserted as needed.

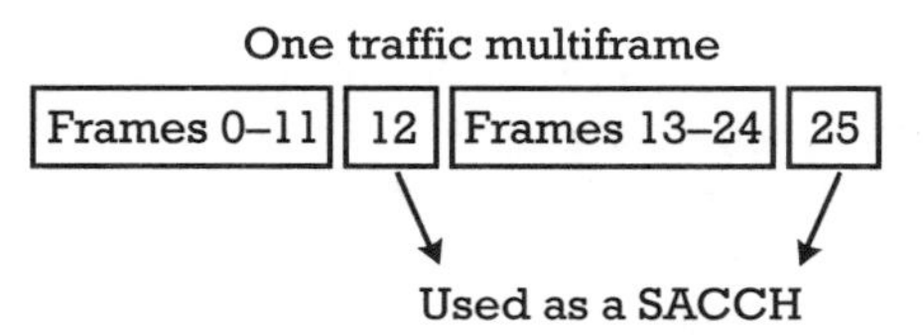

Figure 5.6 The traffic multiframe consists of 26 frames, 24 traffic frames, 1 SACCH frame, (using 12 and 25 alternately) and one not used in full-rate vocoding.

The multiframes are then organized into 6.12-second superframes consisting of 51 traffic multiframes or 26 control multiframes. The final organization is the hyperframe, which consists of 2,048 superframes, and cycles every 3 hours, 28 minutes, and 53.76 seconds.

5.7 Control Channels

Like AMPS and NA-TDMA, if a mobile station is not up on a call, then it is using a control channel to wait for or set up a call, register, or perform any other task needed in the idle state. There are three sets of logical control channels used for signaling and overhead if a call is not in place: BCCHs, common control channels, and dedicated control channels.

The BCCHs and common control channels are the first control channels a mobile station will scan for when it is turned on in a network (see Figure 5.7). These two logical channels are multiplexed on the 51-frame control multiframe in a set manner. If the mobile station needs to respond or access the network, it can use its common control channel, the RACH.

Broadcast channels include:

- *Frequency correction channel (FCCH):* This burst is the "beacon" of the control channel. The mobile will search for this burst, knowing it is the first logical channel in the control sequence. This burst is actually 148 zeros. In GMSK modulation, this equates to a simple sine wave at a specific frequency (approximately 67 kHz). The mobile station can find this sine wave and adjust itself so that is has the same frequency reference as the network. The FCCH is sent in the first time slot. A mobile station can then count seven time slots and wait for the first slot of the second frame, which will always be the *synchronization channel* (SCH).
- *SCH:* In the frame following the FCCH, the BTS transmits with SCH in time slot 0. The SCH has a unique burst structure as well. It contains an extended

FSBBBBPPPP FSPPPPPPPP FSPPPPPPPP FSPPPPPPPP FSPPPPPPPP I

F = Frequency correction channel (FCCH)
S = Synchronization channel (SCH)
B = Broadcast control channel (BCCH)
P = Paging channel (PCH) and access grant channel (AGCH)
I = Idle frame

Figure 5.7 The broadcast and common control channels repeat the same pattern to allow mobiles to find the information they need.

training sequence that is the same throughout all GSM networks. This training sequence allows the mobile station to get exact synchronization. Also included in the SCH is the BSIC and the current frame location in relation to the hyperframe.

- *BCCH:* The BCCH contains the parameters used by all of the mobiles in the cell to set up and receive calls. This might include access parameters and configurations. One broadcast control channel "segment" occupies four frames per multiframe.

Common control channels include:

- *Paging channel (PCH) and access grant channel (AGCH):* The PCH is used to let a mobile station know that it has a call coming in. The AGCH is used to direct a mobile station to another type of control channel—the *standalone dedicated control channel* (SDCCH)—in order to complete the process of setting up a call and/or transferring information. These two channels share a number of frames within a multiframe. The PCHs can be organized to allow for a sleep mode. The mobile station can be assigned a specific group of PCHs to monitor, waking up just in time to do so.
- *RACH:* This is the reverse channel a mobile station will use to originate a call, send signaling messages when not on a call, acknowledge messages from the BTS and register. The name implies that it is random access. Essentially, all of the reverse slots of the common control channel are eligible to be a RACH. A mobile station can select a slot using a specific protocol and transmit a shortened burst. (This is to make sure it stays in the confines of the time slot. The BTS makes a determination of timing based on this burst and advises the mobile station to adjust as needed). If the RACH has been received successfully by the BTS, the network will direct the mobile station to the SDCCH—a two-way control channel designated just for this type of communication between the mobile station and the network. Unlike NA-TDMA, the RACH actually carries very little information. Instead, the SDCCH serves that purpose. Figure 5.8 show the control channels that a mobile can use in a location update.

The dedicated control channels include:

- *SDCCH:* For each mobile assigned to a SDCCH, this two-way logical channel consists of four time slots (carrying one message) in every multiframe. This makes for rather slow transmission speeds, but it is sufficient for the information that needs to be sent. It is on the SDCCH that most of the access messaging takes place. A rather confusing concept is that the SDCCH also comes with a SACCH in order to relay signaling information. Thus, there is a

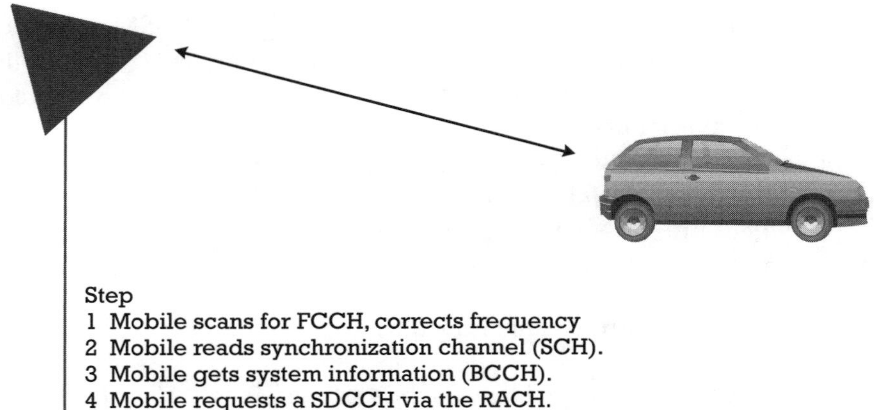

Step
1 Mobile scans for FCCH, corrects frequency
2 Mobile reads synchronization channel (SCH).
3 Mobile gets system information (BCCH).
4 Mobile requests a SDCCH via the RACH.
5 Mobile requests location update (registration) on SDCCH.
6 Mobile is authenticated (SDCCH).
7 TMSI is assigned to mobile (SDCCH).
8 SDCCH is released

Figure 5.8 The mobile station uses the various control channels as it performs a registration (location update).

SACCH multiplexed in the traffic channel multiframe as well as the SDCCH multiframes.

- *SACCH:* This channel carries broadcast messages, power control information, and timing advance in the downlink (forward channel), whereas in the uplink (reverse channel), it has the measurement report, acknowledged power control, and acknowledgment of the timing advance.

5.8 Vocoders and Traffic Channels

Three vocoders are specified in GSM. Currently all three are being installed in newly manufactured mobile stations. The original is LPC-RPE. As in most vocoding processes, speech is divided into 20-ms blocks. In LPC-RPE, each 20-ms block of speech is represented by 260 bits. If you divide 260 bits by 20 ms, you will find the rate to be 13 Kbps.

The vocoding process is very complex. To simplify, some bits of the vocoded stream are considered more important than others and are thus put through more rigorous error correction encoding than others. By adding such encoding and parity bits, the system can tell at the receiving end whether a 20-ms voice stream can be recreated correctly.

This error correction process means that the original 260 bits per 20 ms now turns into 456 bits per 20 ms. If you divide 456 bits by 20 ms, you will get 22,800 bps. This is why you may see the LPC-RPE vocoder called a 13-Kbps vocoder as well as a 22.8-Kbps vocoder.

As mentioned, some of the bits are considered more important than others. If certain bits were missing, it has a much more destructive result on voice quality. In GSM, these bits are divided into classes. Of the original output of the vocoder (260 bits), 50 are considered most important, and are called the class 1a bits. The class 1a bits are given a 3-bit cyclic redundancy code so that errors can be detected when they are received. If there are any errors, the entire frame is discarded. The second class of bits are the class 1b bits. These 132 bits (plus a 4-bit tail sequence), along with the 1a bits, are put through a convolutional encoder, which serves as FEC, adding a degree of protection to these bits considered important.

Thus, 189 bits (53 1a bits, 132 1b bits, and a 4-bit tail sequence, which is used to flush the shift registers in order to initialize them) are put through the convolutional encoding process, so that the output is 378 bits (see Figure 5.9). The remaining 78 bits, called the class II bits, are not put through any protection and are added to the 378 encoded bits for a total of 456 bits per frame.

Two newer vocoders are specified. The first is the EFR vocoder, which was developed in response to complaints about the original vocoder's voice quality. EFR

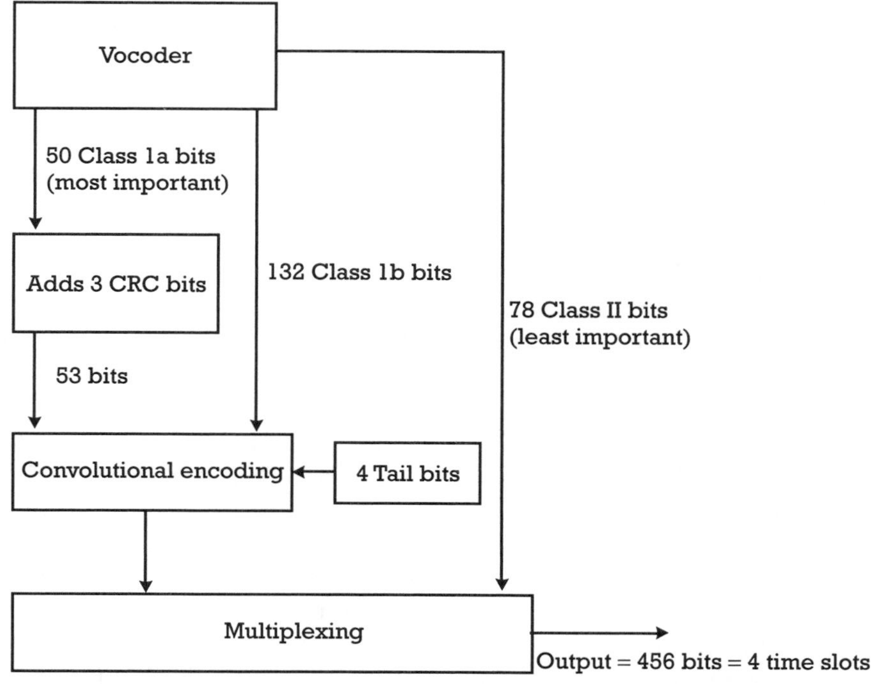

Figure 5.9 The process of taking 189 bits out of the vocoder and error detection and then correction encoding some of them to produce 456 bits: 20 ms of speech becomes four time slots of data.

uses the same ACELP technology used in NA-TDMA. The second is the half-rate vocoder. It operates at half the rate of original vocoder, which allows it to fit twice as many users in the same amount of space.

As mentioned earlier, built into every traffic multiframe is a SACCH, which is used for signaling information transfer during a call. If the SACCH is not fast enough, the system can use a FACCH. In this case, the FACCH will steal time from the voice channels. A FACCH will be spread over eight slots spread out over eight frames (same as with TCH block of 456 bits, which is also spread over eight slots or frames).

5.9 Messaging

For those without a good background in data, the messaging process in GSM can be a bit confusing. To oversimplify, there are three types of messages in GSM: information, supervisory, and unnumbered. Supervisory and unnumbered messages simply control the flow of information messages. If you are familiar with ISDN, this might sound familiar. In any event, the information messages are important to understanding the overall air interface.

There are three sets of information messages: call management, mobility management, and radio resources management.

The radio resource management messages include the BCCH messages, which contain all of the setup parameters for the cell. Also included in this group is the CHANNEL REQUEST message, which is sent on the RACH. The radio resources management group also contains the messages associated with the PCH and AGCH (i.e., the process used to assign a mobile station to a SDDCH in order to begin to set up a call). Other messages include:

- MEASUREMENT REPORT: This message is sent on the SACCH and is the primary purpose of the SACCH, whether it is associated with the SDCCH or a traffic channel. The measurement report is used to send power control measurements.
- HANDOVER COMMAND and ASSIGNMENT COMMAND: These are used to perform handoffs (handoffs are called *handovers* in GSM). The two commands perform the same task, moving the mobile station from one channel to another. The key difference is that when the network uses the HANDOVER COMMAND message, the mobile will move to the new channel and transmit a shortened burst containing the HANDOVER ACCESS message. This shortened burst and message is used to adjust the timing of the mobile station, which is controlled by the BTS in a PHYSICAL INFORMATION message. If the ASSIGNMENT COMMAND is used, the mobile station does not go through this timing alignment process. These messages are placed in a FACCH.

Mobility management messages pertain primarily to registrations and authentication procedures. These messages take place in the SDCCH. The mobile station tells the network it is in the neighborhood via the LOCATION UPDATING REQUEST, and it is confirmed by the response from the network with either a LOCATION UPDATING ACCEPT or REJECT message. When a mobile station turns off, it can also send an IMSI DETACH INDICATION, which essentially tells the network not to look for it anymore.

Authentication in GSM is similar to other formats. Ki, a secret key on the SIM card, is used in conjunction with a random number sent by the BTS in a complex authentication algorithm. The result of this function is then sent to the BTS for verification (the network has the Ki stored as well).

The network can obtain the pertinent information from the mobile station as part of the authentication process via IDENTITY REQUEST and IDENTITY RESPONSE messages. Here, the IMSI, IMEI, and TMSI will all be sent by the mobile station (thus, if a SIM or phone was reported as stolen, the network can find out and consequently reject it).

The last set of messages is the call management messages. These messages are sent on either the SDCCH before a call is set up or the FACCH after the call is established. These are the messages in which the mobile station originates calls, confirms a traffic channel, gets the alert message, releases a call, sends and receives DTMF functions, and performs other such tasks during a call.

5.10 Burst Types

As has been mentioned, several of the channels use different types of bursts to convey information or help adjust parameters. The time slot itself is 576 ms, or 156.25 bits. GSM uses five different bursts to fill this time slot (see Figure 5.10):

- *Normal burst (uplink + downlink):* This standard burst will fill the slot completely. It requires accurate timing advance in the uplink or it can interfere with neighboring time slots.
- *Frequency correction burst (downlink):* This is the FCCH, essentially a sine wave that equates to 148 zeros.
- *Synchronization burst (downlink):* This SCH contains a longer training sequence, which allows the mobile receiver to get synchronized. It also contains the BSIC and the frame number in relation to the hyperframe.
- *Access burst (uplink):* This is used for access on the RACH and during a call in order to correct timing. It is shortened to ensure that it fits in the time slot regardless of severe time alignment problems.

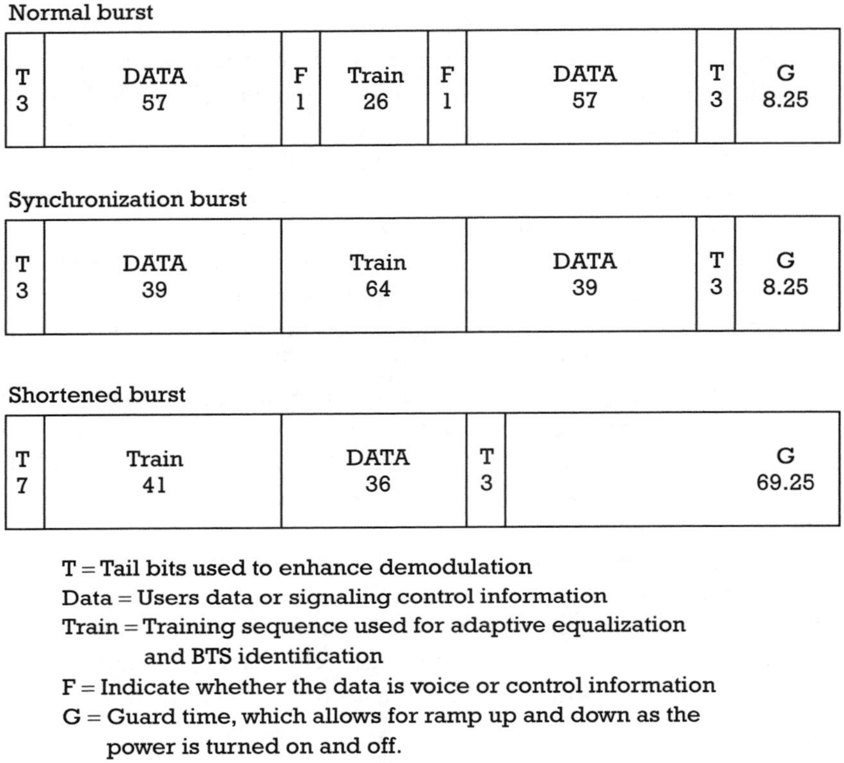

Normal burst

T 3	DATA 57	F 1	Train 26	F 1	DATA 57	T 3	G 8.25

Synchronization burst

T 3	DATA 39	Train 64	DATA 39	T 3	G 8.25

Shortened burst

T 7	Train 41	DATA 36	T 3	G 69.25

T = Tail bits used to enhance demodulation
Data = Users data or signaling control information
Train = Training sequence used for adaptive equalization and BTS identification
F = Indicate whether the data is voice or control information
G = Guard time, which allows for ramp up and down as the power is turned on and off.

Figure 5.10 The three primary burst structures used in GSM. Not shown is the FCCH, which is really just a sine wave, and the dummy burst, which is used as a filler.

- *Dummy burst (downlink):* This is simply used as a filler for unused physical channels (i.e., time slots).

5.11 Frequency Hopping

As discussed in Chapter 1, one of the best ways to fight interference is to spread the signal out. CDMA does this by spreading a signal with extra chips. GSM can do this by hopping frequencies. At every frame, the system can hop to a new frequency, thus spreading the signal out. This serves the purpose of fighting frequency-specific fades.

It also helps fight cochannel interference, as it keeps hopping from channel to channel and only stays on one particular channel for a fraction of the time. Thus, if two mobiles are assigned the same channel in their frequency-hopping pattern, they will seldom use that same frequency at the same time—and even if they do meet, it is only for one frame.

5.12 Handovers

In GSM there are four type of handoffs:

- Handoffs between time slots or frequencies in the same cell (intra-BTS);
- Handoffs from BTS to BTS under control of the same BSC (inter-BTS);
- BTS under one BSC to BTS of another BSC, but all under the same MSC (inter-BSC);
- BTS in one network to BTS of another network, under different MSCs (inter-MSC).

Handoffs in GSM are similar to handoffs in NA-TDMA in that the mobile station assists the network in making handoff decisions by measuring neighboring base stations in between its assigned burst slots. The mobile stations determine the BER of its received signal and a related receive quality measurement, and they report this to the BSC via the MEASUREMENT REPORT on the SACCH. Because the BTS it may want to handoff to will probably be a different distance away than the current serving BTS, often the timing advance will need to be recomputed. Thus, as described in Section 5.9, the first step after moving to a new BTS is to transmit a shortened burst message called the HANDOVER ACCESS message. The BTS computes the timing advance and relays it back to the mobile station via a PHYSICAL INFORMATION message. The HANDOVER COMPLETE message, sent on the FACCH by the mobile station, completes the process and the conversation continues.

5.13 Multipath Equalization

A term often discussed in GSM is *multipath equalization*. When you have read through the CDMA chapter, you will learn about the Rake receiver, which demodulates multipath signals separately and then combines them coherently. GSM (as well as NA-TDMA) also must deal with multipaths. If you look at the slot structure of the GSM burst, you will find a training sequence in each burst. This known sequence can be analyzed by the receiver and used to suppress the distortion caused by multipath propagation. This is a fairly complex process and beyond the scope of this text.

5.14 Stage 2 and Beyond

GSM was rolled out one stage at a time. Stage 1 included much of what was discussed in this chapter. Stages 2 and 2+ include such items as an EFR vocoder for better voice quality and higher speed packet data services.

The high-speed data applications will probably drive the growth of wireless formats. It can be argued that the upgrade path for GSM in the future, similarly to CDMA, will cover many of the features most wanted in a next generation (termed 3G) system, without a complete overhaul.

The first emerging data format is called *high-speed circuit-switched data* (HSCSD). The term *circuit switched* means that there is an open data channel at all times between the user and the network. This can become costly. With HSCSD, data rates of up to 56 Kbps can be achieved. This is performed by allocating multiple time slots to a single user.

The second emerging data format is GPRS. This is a packet-data system, which means data is sent in packets to the user as needed and does not require a dedicated circuit. This can be cost effective, as a user would normally pay only for the amount of data actually transmitted. With GPRS, data rates could get to 144 Kbps, although most users would probably use standard rates around 56 Kbps. A third type of data transmission called EDGE is also on the horizon. EDGE data rates might triple the rates achievable with GPRS. The third generation formats being discussed consider 2.08 Mbps to be the goal. However, as you can see, an upgrade path for current hardware could approach that type of transmission rate. These formats are discussed in detail in Chapter 9.

5.15 Conclusion

GSM has become closer to being the world standard for wireless communications than any other system. With networks in virtually all geographic regions of the world, and substantial growth continuing, the GSM standard still continues to evolve with the development of high-speed data applications. In addition, most NA-TDMA operators are now transitioning to GSM. While CDMA may grow at substantial rates over the next few years, it seems impossible to think they could ever match the current penetration of GSM. GSM is used extensively in 900-MHz, 1,800-MHz, and 1,900-MHz bands, and now in the 800-MHz band as well. As the technology matures even further, and the cost for the technology drops even lower, GSM will certainly still continue to grow.

CHAPTER 6

GPRS and EDGE

6.1 Introduction

By far, GSM is the world's most popular cellular system. Almost every country in the world has a GSM system operating, including ubiquitous coverage throughout Europe and most of North America. In addition, China, with the largest population in the world, also has a substantial GSM network. If you have read the earlier chapters on GSM, you will remember that GSM was designed specifically for voice communications, with excellent roaming capabilities. It was not originally designed to have an easy upgrade path to data transmission.

HSCSD was an implementation of circuit-switched data on GSM channels, but the lower speeds and circuit-switched nature of the option do not make it a viable option for true high-speed data applications. In addition, operators who implemented NA-TDMA networks also were without an upgrade path that gave them a high-speed option. Most of these operators are migrating to GSM, but they will also need a solution to offer their customers access to high-speed data transmissions. As a result, two new formats were designed to allow for an easier integration into the GSM network. Both have 200-kHz bandwidth signals, which allow them to be seamlessly added to the spectrum, as GSM also has a 200-kHz bandwidth.

GPRS will more than likely be the standard for high-speed data service for GSM and NA-TDMA operators for some time—at least until WCDMA networks are built out substantially enough (see Figure 6.1). Enhanced Data for GSM Environments, now changed to Enhanced Data for Global Evolution, but better known as EDGE, is a next step to higher data rates. It uses higher order modulation techniques to deliver significantly higher data rates in the same 200-kHz channel as GSM.

Circuit-switched data, such as HSCSD in GSM, can serve the basic needs of mobile users as far as data goes. In GSM, it is possible to provide 14.4 Kbps of data per time slot. HSCSD enabled the combining of time slots to get close to 64 Kbps per user. The problem, though, is when a circuit-switched connection is needed, it is dedicated to the user and remains on, much like the user on a voice call. This uses up the network's capacity and resources. Data applications, unlike voice, do not need to be real time and are often bursty in nature—hence, a circuit-switched connection

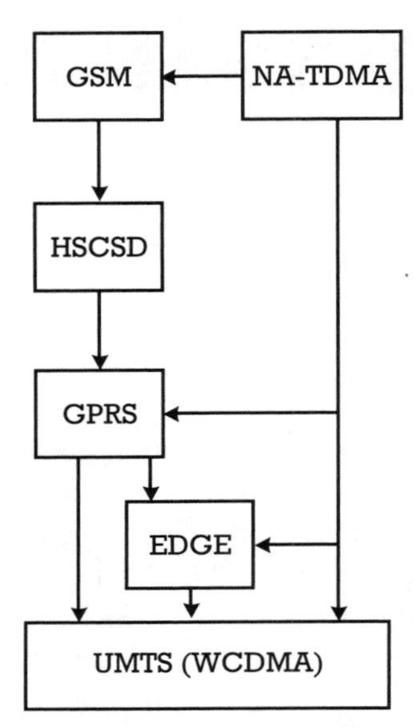

Figure 6.1 The apparent technology evolution path for GSM and NA-TDMA operators would seem to go through GPRS, and possibly EDGE, although some operators may opt directly for Universal Mobile Telecommunications System (UMTS).

is using resources it really does not need. A better method for delivering data is through packet switching—where the physical channel is shared, and data can be sent to and from the terminal as required. This can allow for an always-available data connection, as opposed to having to establish the circuit every time a user wants to send or receive data.

In addition, the data world is an Internet-based world. Thus, the core network needs to be *Internet protocol* (IP) based as far as its architecture. Once an operator has this IP-based architecture in place, it can be used for additional growth as the technology is implemented. For instance, 3G formats can utilize this IP-based core network without the need for additional investments. You will find WCDMA makes use of GPRS architecture for its data services. Hence, once GPRS is deployed, it allows for a data and an IP evolutional path.

In general GPRS is designed to offer a cost-effective high-speed data solution to GSM operators. The EDGE format will work in a very similar manner as GPRS. However, by utilizing different modulation techniques, it will be able to achieve higher data rates. In addition, by using voice-over-IP techniques, it may be possible for EDGE radios to support not only high-speed data, but voice transmissions as well. It should be added that GPRS, or any high-speed data format, can support voice-over-IP technology, though it may be more effective with EDGE. Although no operators have committed to this, several are considering the possibilities.

6.2 GPRS Basics

A GPRS mobile unit works remarkably similar to a GSM mobile. A network is divided into routing areas, which are simply clusters of cells. The mobile scans all of the cells during idle periods, constantly searching for the best cell to receive and transmit to, switching to the best cell as needed. If the mobile needs to be contacted or needs to establish a connection, the paging process also works in a very similar manner as the GSM voice calls. In the backhaul, however, things are a bit different. The BSC will route the data call through a GPRS support node. In the *GPRS public land mobile network* (PLMN), two of the support nodes that form the GPRS backbone will be referred to: the *serving GPRS support node* (SGSN) and the *gateway GPRS support node* (GGSN) (see Figure 6.2). The former is the gateway to data services from within the mobile network; the latter is the gateway from the PLMN to the outside IP world.

Once data service is assigned to a mobile, it is the job of the SGSN to track the location of that mobile within the network and ensure that the mobile is authenticated and is receiving the correct level of quality of service. It is the job of the GGSN to interface with the outside data world. This is all done independent of the RF interface, as it is on the backhaul side of the BSC.

6.3 Packet Data Protocol Basics

This text is intended to focus on the RF side of the wireless formats. However, when discussing data operation, it is useful to have a basic understanding of *packet-data protocol* (PDP), which are the network layers.

There are two network layer protocols that GPRS supports: X.25 and IP. Neither will be discussed in detail in this text, except to say that both protocols assign

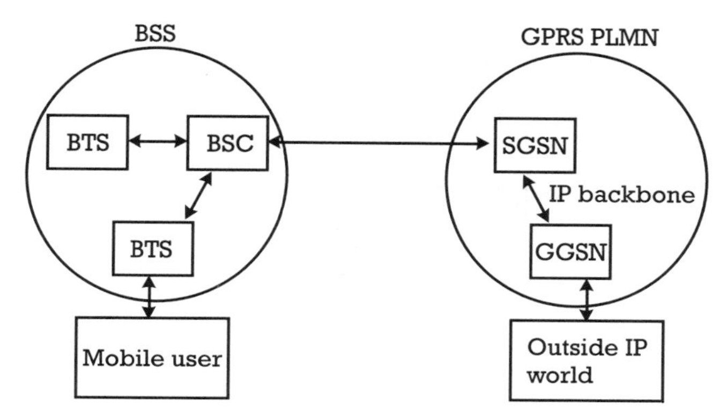

Figure 6.2 The GPRS backbone will be used for EDGE and WCDMA. The two primary gateways are the SGSN, the gateway to the mobile network, and the GGSN, the gateway to the outside IP world.

network addresses (or PDP addresses) to the devices in the network in order to route data correctly. In GPRS, this means the GGSN will obtain a PDP address from the public packet data network. The network can then, through the SGSN, can route data to specific mobiles, which are assigned their own PDP address by the GPRS PLMN. Thus, the GGSN will appear to the public packet network as a normal data gateway, hiding the fact that the users are actually mobiles. This gateway can also serve the role of firewall as well, which hides the mobile from the external network so intruders cannot locate and attach the mobile.

Thus, in order for a mobile in a network to receive and send data, it must first attach itself to the SGSN and then activate its PDP address. This activation process involves the mobile being assigned a PDP address by a GGSN, which is associated with the SGSN. Data technicians often call the record of these associations the PDP context. While a mobile can only attach to one SGSN at a time, it can actually receive data from multiple GGSNs via multiple PDP addresses. This addressing is shown in Figure 6.3.

6.4 GPRS Phones and Devices

GPRS terminals are classified depending on their capabilities. Currently three types of classifications are used:

- Class A mobile stations can make and receive calls on GSM and GPRS at the same time.

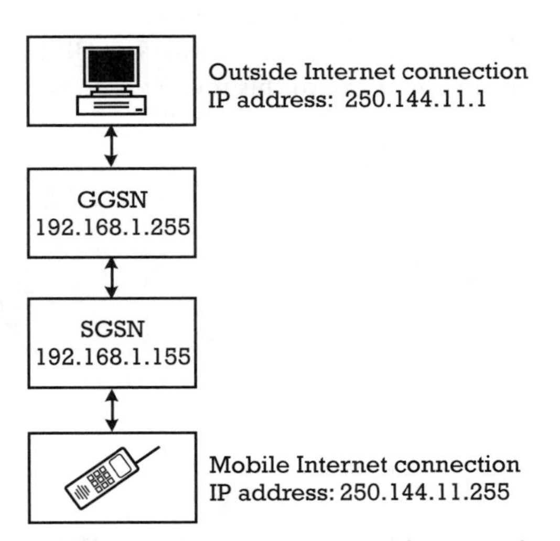

Figure 6.3 In the IP data world, addresses are used to identify nodes. The mobile phone becomes an IP address.

- Class B mobile stations can make and receive calls on GSM and GPRS, but not at the same time.
- Class C mobile stations need to be manually selected as far as whether they will operate in GPRS or GSM modes. This selection is done at the time of subscription of service by the user; thus, when a customer purchases a class C mobile, he or she must select either GPRS or GSM mode.

In addition, as you will see in later sections, GPRS makes use of multiple time slots in order to increase the data rates. You will remember that a GSM carrier has eight time slots. Different multislot classes of GPRS devices will be able to use different combinations of transmit and receive slots in order to offer different data rates, on the uplink as well as downlink. There are 29 multislot classes, with class 1 allowing one receive and one transmit slot, while class 29 allows for use of all eight transmit and all eight receive slots. The classes in the middle allow for various configurations of asynchronous data transmission (i.e., the uplink and downlink having different data rates).

GPRS and GSM assignment of slots, as shown in Figure 6.4, is just an example. Any slot can be assigned to GSM or GPRS, unless it is preassigned permanently, which is not very efficient. A dynamic assignment is best. However, hybrid assignments enable some slots to be assigned to GSM and some slots to GPRS, and then based on the need one can be assigned to GSM or GPRS. Voice always takes precedence; thus, GSM should have the first choice.

6.5 Understanding the Layers

While this text focuses on the RF links between the base stations and mobile stations, it is important to understand the interaction of the layers of data in order to comprehend the coding schemes and packet transfer processes.

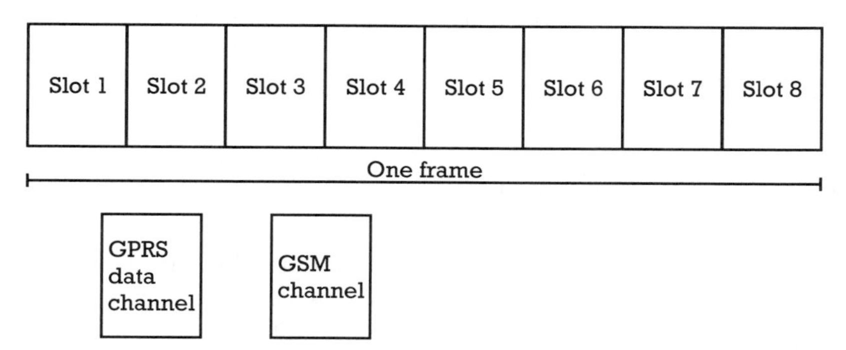

Figure 6.4 GPRS uses the same time slot structure as GSM. In fact, GPRS slots can be intermingled with GSM voice slots.

The RF layer, layer 1, manages the physical link between the mobile and the base station. This link is divided into two sublayers in order to show the various functions:

- *Physical RF layer:* This includes modulation and demodulation (in GPRS, this would be GMSK; in EDGE, it would be 8PSK).
- *Physical link layer:* This manages the information needed for the physical channel, such as the error correction, interleaving, measurement reports, power control, and cell site selection (note that handoffs in the traditional sense do not take place in GPRS; sites to transmit to or from are selected prior to packet transmission).

The next layers up are the *radio link control* (RLC) and *medium access control* (MAC) layers, which will supervise the logical links between the mobile and the base station system and the mobile station. These layers control radio link access, and they contain the various logical channels which get data to and from the mobile from and to the BSS. They are also the layers that will format data frames from and to the next layer up, the *logical link layer* (LLC). These links interface with the frame relay network, which is used to link the components of the core network in GPRS. The *BSS GPRS protocol* (BSSGP) is a layer that specifically links the LLC with the BSC. Figure 6.5 shows the *protocol stack* for GPRS.

The LLC handles the interfacing between the SGSN, BSS, and mobile. LLC packets contain the actual user data. It is a ciphered link between the mobile station and the SGSN (via the BSS). It is in this link that authentication takes place. You might recall in GSM that ciphering/deciphering took place in the BSS—in GPRS, it will take place at the SGSN at this layer. The mobile is attached to the SGSN through the LLC layer. It has its own error detection. It is addressed via an identifier called the *temporary logical link identifier* (TLLI). If you remember from GSM, the identifier for voice is the IMSI. The TLLI is much like the IMSI in GSM in that it uniquely identifies the mobile—in this case, for the LLC layer. It is this layer that allows for easy convergence of wireless technologies, as the LLC packets would be standardized for the various 3G formats. Hence, the SGSN should be interoperable with GPRS, EDGE, UMTS, and perhaps even CDMA2000.

6.6 GPRS Physical Channels

As stated earlier, GPRS uses the same modulation and radio frame structure as GSM. This allows GPRS channels to be assigned dynamically by the base station depending on the demand. Thus, within a TDMA frame, a few of the slots can be designated for GPRS physical channels, while the remaining can be used for voice or

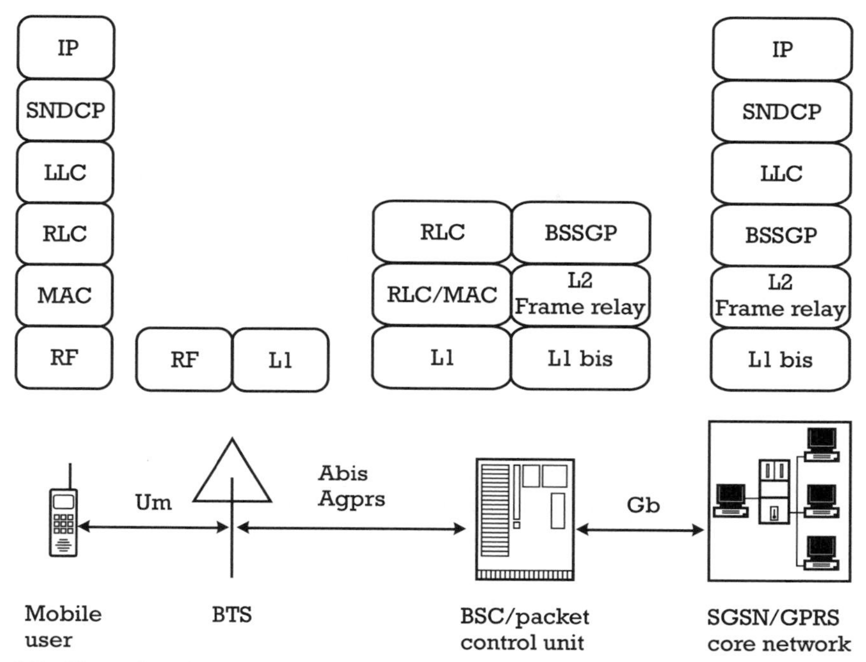

Figure 6.5 The various layers at the four main components of a GPRS network—the mobile, the BTS, the BSC/packet control unit (PCU), and the core network.

other physical channels. For a review of GSM physical channels and call processing, the reader can refer to Chapter 5.

For GPRS operation, there is one new physical channel that will be used, called the *packet data channel* (PDCH). Figure 6.6 shows the GPRS slot (burst) structure. Remember, the GPRS slot will be identical to the GSM voice slot in terms of physical attributes; that is, it has the same power profile (burst), will require timing advance to compensate for being different distances from the base station, and uses the same modulation. Each four bursts will carry the data of one RLC block of coded data. This will be discussed in more detail later.

There are 52 frames in a multiframe, which is how time slots are scheduled in GPRS. Figure 6.7 shows this multiframe structure. You will notice that each frame

Tail bits	Encrypted data	Training sequence	Encrypted data	Tail bits	Guard period
3 Bits	57 Bits + 1-bit coding scheme flag	26 Bits	57 Bits + 1-bit coding scheme flag	3 Bits	8.25 Bits

Figure 6.6 The physical burst structure of a GPRS burst.

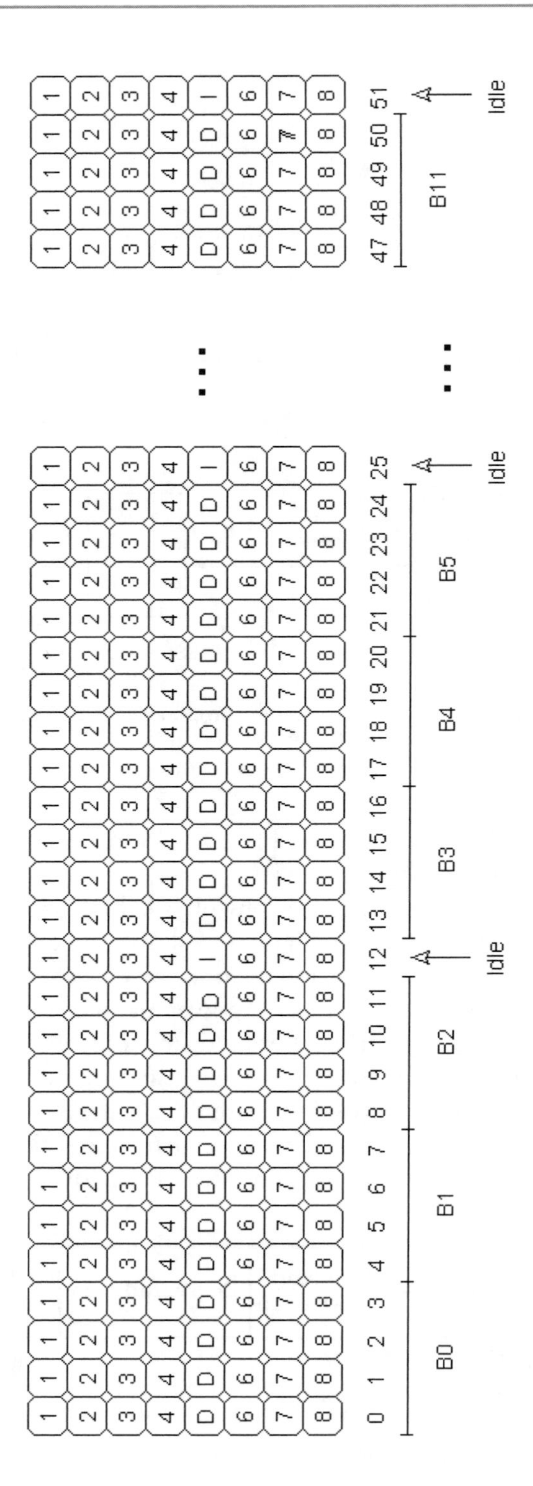

Figure 6.7 The 52-frame multiframe is used to divide the frames into data (D) blocks, and set aside specific frames for idle time (I)—which are used for cell measurements and timing control.

contains eight slots, all of which last 4.62 ms, just as in GSM. Also, as in GSM, each slot is then 0.577 ms. The 52 frames that make up the multiframe, therefore, take 240 ms to complete. The base station will assign PDCH physical channels to a particular time slots, in this case to slot 5. At times, the PDCH channel will be idle, allowing the mobile to measure neighbor base stations for handoff purposes. Also, the slot may be used by the base station and mobile to determine time delay, which will require a specific logical channel known as the *packet timing advance control channel* (PTACT) to be sent in the slot on the downlink and uplink. The rest of the time, GPRS data will be sent in the slots.

The 52-frame multiframe is further divided into 12 blocks, shown in Figure 6.7, which exclude the frames used for timing and idle. The blocks each contain four frames. You should remember that four bursts are used for each RLC/MAC block of coded data.

6.7 GPRS Coding

In voice communications, it is important that there not be long delays in reception of data, because a voice conversation takes place in real time. Data, on the other hand, does not always require this real-time reception. Thus, a transmitter having to resend packets of data because the first transmission was not received is often quite acceptable and not even noticed by the user. Some data applications, however, do require more real-time transmissions and cannot tolerate too many retransmissions—such as video conferencing or games.

It might be useful to go over the definitions of data rate versus throughput. Data rate is the rate at which data is being sent, regardless of how many errors might be felt at reception. Throughput is the measurement of actual received data. Thus, it is possible to have a very high data rate, yet a very low throughput, if much of the data needed to be resent.

As a wireless operator, a goal would be to try and offer constant throughput in as ubiquitous a manner as possible. Considering the different RF environments, this is a major challenge, as some areas might have high interference while others have none, and some users will be close to the cell site while others are at the fringe. Hence, GPRS attempts to mitigate this issue with the use of variable error protection coding formats. When the RF environment is good, a coding scheme is used that allows for maximum throughput by using little protection. When the RF environment is bad, more thorough coding is used, which protects the bits more. This results in an optimized throughput, though at the expense transmitted data (see Figure 6.8).

Thus, GPRS has four coding schemes that can be used, depending on the situation. These schemes are designated CS-1 to CS-4. You should recall how convolutional encoders are characterized, that is, a rate of 1/2 implies for every one unit input into the encoder, 2 bits will come out.

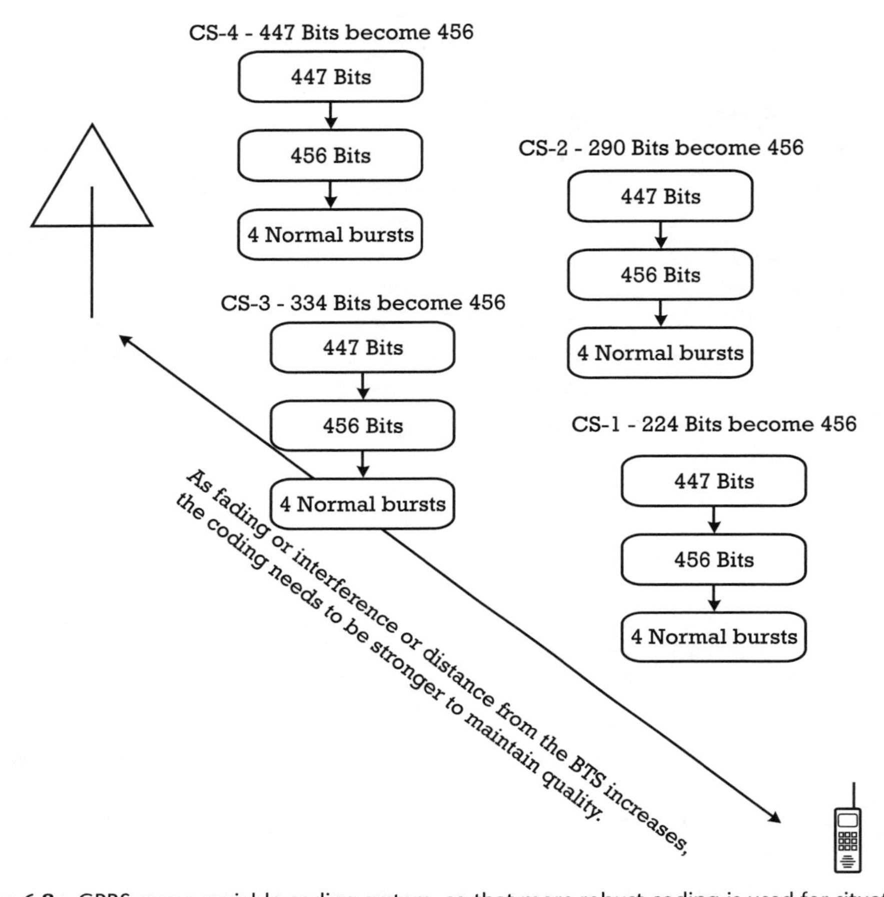

Figure 6.8 GPRS uses a variable coding system, so that more robust coding is used for situations where the RF will be weak, and less where it is optimal—this allows for optimal data throughput.

- CS-1 provides for the most correction and detection for situations where the carrier-to-interference levels are very low (low carrier and/or high interference). In situations like this, the data needs as much help as possible so that it will not continually need to be repeated. CS-1 uses a 1/2 code rate, which will result in a 9.05 Kbps payload data rate, or 181 bits per 20 ms. In a 1/2 encoder, for every 12 bits in, 24 bits come out.
- CS-2 uses a 2/3 code rate that yields a payload data rate of 13.4 Kbps, or 268 bits per 20 ms. In a 2/3 encoder, for every 12 bits in, 18 bits come out.
- CS-3 uses a 3/4 coder that yields a payload data rate of 15.6 Kbps, or 312 bits per 20 ms. In a 3/4 encoder, for every 12 bits in, 16 bits come out.
- CS-4 is used in ideal situations where carrier-to-interference levels are optimum. Here, no correction encoding is used at all, so the code rate is 1 (12 in,

12 out) for a maximum payload data rate of 21.4 Kbps, or 428 bits per 20 ms. Hence, the maximum data rate for a GPRS frame, assuming all slots are being used for data, is $8 \times 21.4K = 171.2$ Kbps. But you can see that is would only be in the most ideal locations and situations, as there is no error correction at all.

GPRS also uses various interleaving techniques, just as GSM voice used, although the techniques are slightly different. This text will not delve into the details of the interleaving process; however, it is important that the technician understand the concept. Interleaving, you will remember, is used to spread bits out. Thus, if a burst has significant errors, it does not affect adjacent bits, which would make it impossible to correct. Instead, interleaving causes it to have rather random bits across several bursts, thus limiting the impact and allowing the error correction to work.

6.8 Bursts

Section 6.6 described the physical slot structure GPRS uses for each burst. As discussed in Chapter 5 on GSM, data is encoded such that the output will be a 456-bit packet every 20 ms as seen in Figure 6.9. GPRS uses these same parameters to encode its data. Thus, as mentioned earlier, depending on the coding scheme used, the packet that is to be transmitted can have from 181 bits of payload data up to 428 bits of payload data in each burst of 456 bits. You can see more clearly at this point why coding scheme 4 is unrealistic in most cases, as the correction encoding will be so weak.

Because each burst is 0.577 ms and carries two blocks of 57 information bits each, as seen in Figure 6.6, it requires four bursts to transmit each 20-ms block of encoded user data. Each four bursts, therefore, carries 456 bits of encoded data. This encoded data contains either 181, 268, 312, or 428 bits of actual payload

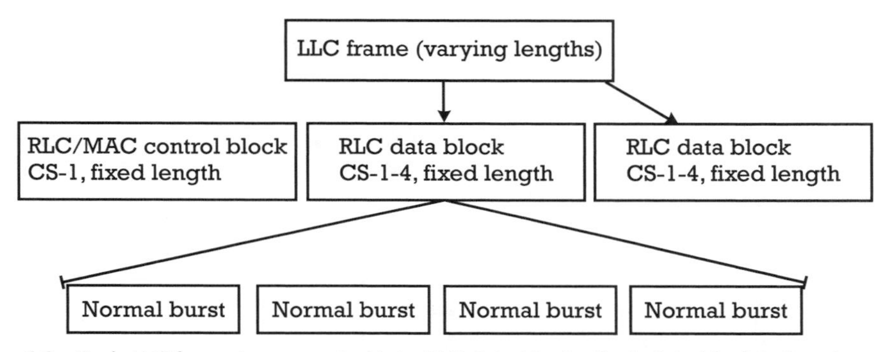

Figure 6.9 Each LLC frame is segmented into RLC data blocks. Each data block is then transmitted over four normal bursts.

information, depending on whether coding scheme 1, 2, 3, or 4 is used, respectively. Thus, maximum throughputs will depend on the coding scheme used in addition to how many slots are used per frame, in further addition to whether the frames are received correctly and needs to be repeated or not. As will be discussed in later chapters under test and measurement, it is important to understand how the quality of the signal can now drastically affect capacity, as frames that need to be repeated will tie up slots.

In GPRS, just as in GSM, there is also a shortened burst that carries the logical RACH—in this case, the *packet random access channel* (PRACH). See Chapter 5 for more details on the access burst.

It is important to understand that the bursts are the last time the data is manipulated, taking place in the RLC/MAC layers. Remember that the LLC (i.e., the logical connection layer where messages can be transported between the SGSN, BSS, and mobile) uses frames that are variable length, but in the RLC/MAC, which takes the LLC frames and manipulates them for radio transmission, the frames are formatted into several set-length blocks. Using header information, the system is able to put the original LLC frames back together after they are received.

6.9 Channel Allocation

In other formats, you will remember that there are usually several different physical channels used to send and receive various control information. As you have seen earlier, GPRS uses one physical channel, the PDCH. This PDCH carries packets of data that have been coded and assigned to bursts to be transmitted. Aside from the GSM SCH and FCCH, GPRS uses its own set of logical channels, mapped on the PDCH.

You might recall that the benefit of packet-switched data versus circuit-switched data is that the channel does not get assigned to one individual user, but rather the packets get assigned to all of the users, who share the channels. In this way, it is an always-available data channel but does not hog the physical channel resource—thus, the concept of packet MAC, which describes the process of messaging in the form of packets. In this way, packets can be assigned based on the precise moment's needs.

GPRS will use three types of MAC modes to control transmissions from the mobile: fixed allocation, dynamic allocation, and extended dynamic allocation (see Figure 6.10).

- *Fixed allocation:* If a mobile application requires a consistent data rate, this mode can be used. It assigns a set of PDCHs for a fixed amount of time. Because the mobile is assigned the channel, it does not need to monitor the uplink for availability. Rather, it can transmit and receive freely. For

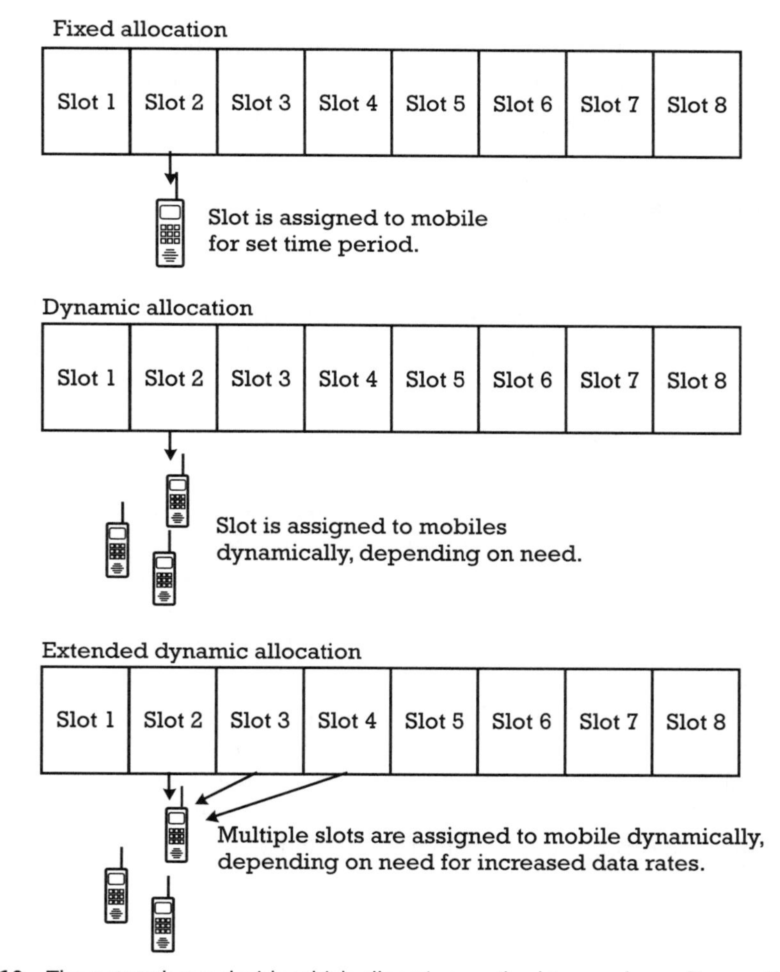

Figure 6.10 The network can decide which allocation method to use depending on the end user's application.

applications that demand real-time constant speeds, such as videoconferencing, this mode is very useful.

- *Dynamic allocation:* This mode allows the network to assign time slots to a mobile as it needs them. Each GPRS time slot can have up to eight mobiles assigned to it. The mobile knows it can transmit in the uplink when it recognizes an identifier assigned to it called the *uplink status flag* (USF). When the USF matches its own, the uplink slot is freed for it to transmit in.
- *Extended dynamic allocation:* You will remember that to increase data rates, GPRS allows the use of more than one time slot per frame. The extended

dynamic allocations allow for this by using the USF to let the mobile know it can transmit, except now the mobile can transmit in multiple time slots as opposed to just the one matching the downlink channel.

6.10 Logical Channels

Just as in GSM, the GPRS logical channels can be grouped depending on whether they are common (i.e., many mobiles will use the same channels for information, access, and paging purposes) or whether they are dedicated (i.e., the channel is specifically dedicated to one mobile at a given moment) (see Figure 6.11). As mentioned earlier, there is one physical channel, the PDCH, onto which these logical channels are mapped. The GPRS format does make use of the SCH and the FCCH, which are part of the GSM system for initial frequency tuning and synchronization. You should review the sections in the GSM chapter on these channels.

The logical channels are also classified into four groups of channels: packet traffic channels, packet dedicated common control channels, *packet broadcast control channels* (PBCCHs), and *packet common control channels* (PCCCHs) (although the dedicated channels can be associated as traffic channels because they are used during traffic states).

The PBCCH is a downlink-only channel that acts similar to the BCCH in GSM. In fact, the mobile will learn of the PBCCH from the BCCH. The BCCH will provide the time slot number for the PBCCH, the training sequence code for the PBCCH, as well as the RF channel number. It is possible to configure GPRS without a PBCCH, putting the broadcast information actually in the BCCH. The PBCCH then

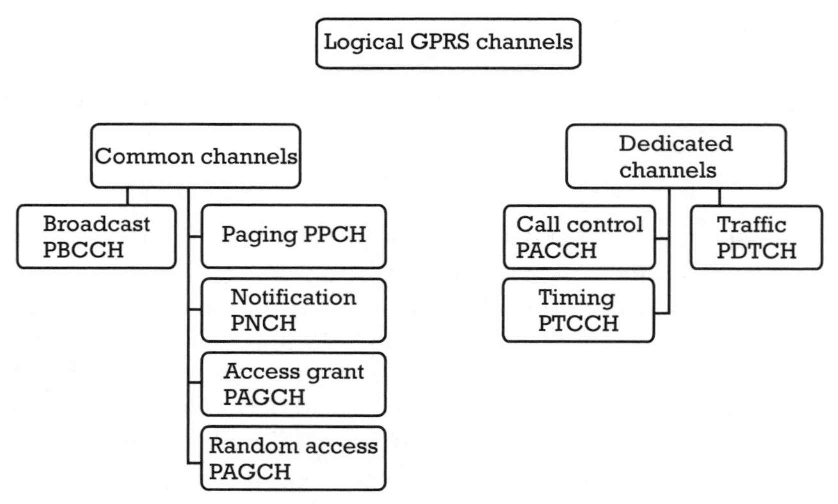

Figure 6.11 There are four sets of logical channels: broadcast, common control, dedicated control, and traffic channels.

broadcasts general information needed to set up GPRS mode, such as power control parameters, operation modes, access methods, network control-order parameters, and information regarding the GPRS channels on the cell site transmitting as well as bordering sites.

The PCCH channels are used just as the CCCH in GSM. They contain the signaling needed to transfer the packet data. There are several logical channel functions included under the PCCCHs:

- *Packet paging channel (PPCH DL):* This downlink channel is used to get the mobile ready to receive data. Identical to the PCH in GSM, the PPCH DL uses paging groups to find the mobile. Note that this channel is for control signaling prior to data call setup. Once the call begins, control signaling will take place on the PAACH.
- *Packet access grant channel (PAGCH DL):* This downlink channel sends the resource assignment message, which assigns the mobile a traffic channel. Oversimplifying a bit for comprehension, in a mobile-terminated data call, the mobile will receive the PPCH DL, which lets it know it has a call and then will receive the PAGCH DL to get the traffic channel assignment.
- *Packet notification channel (PNCH DL):* This downlink channel is used for point-to-multipoint multicast (PTM-M). It notifies the mobile that it has such traffic incoming. This is broadcast traffic intended for a large number of mobiles.
- *Packet random access channel (PRACH UL):* This uplink only channel allows the mobile to initiate and uplink transfer of data or signaling via a shortened burst. There are two types of PRACH bursts: an 8-information-bit standard packet random access burst, or an 11-information-bit extended packet random access burst. The access burst methods are the same as in GSM, allowing for timing advance setup as well. The difference is the 11-information-bit burst, which adds extra bits that are used for priority setting.

During the actual traffic call, there are two addition control channels that are dedicated to the mobile:

- *Packet associated control channel (PACCH UL or DL):* These uplink and downlink channels are used for signaling during a call. This includes resource management (channel assignments) as well as power control and acknowledgments of received messages.
- *Packet timing advance common control channel (PTCCH UL or DL):* This channel is used to ensure the timing advance is optimized (see Figure 6.12). You will remember from earlier chapters on TDMA that it is critical that timing of the slot transmissions be monitored and adjusted in order to ensure that

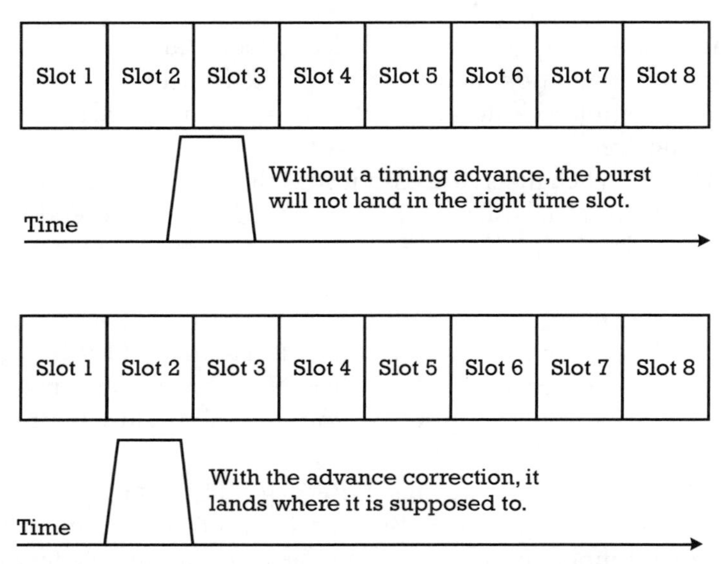

Figure 6.12 The timing advance ensures the bursts arrive at their destination in the appropriate time slot.

they reach the receiver in the appropriate time slot. The mobile transmits a random access burst in the uplink, and the BTS will make measurements, sending an adjustment on the downlink PTCCH. Timing advance is measured in bits and can be adjusted up to 63 bits, with each bit lasting 3.69 ms. More on timing advance can be found in earlier chapters.

The packet traffic channel (PTCH UL or DL) is used in the uplink or downlink to transmit the user data. Up to eight PTCHs can be assigned to one mobile at a time. In the case of PTM-M, up to eight channels can also be assigned to multiple mobiles.

6.11 Overview of GPRS Processes

There are three states of operation for the GPRS system: initialization/idle, standby, and ready states.

Initialization naturally involves the mobile establishing itself on the network. This involves receiving the broadcast channel, from which it can determine the correct frequency and time slot to monitor for packet traffic. In the 51-frame multiframe, time slot 0 contains the broadcast channel.

Just as in voice communications, a mobile must first perform the registration process. In GPRS, this is called the *location update,* allowing the network to know what cell the mobile is in. Once the mobile gets the initial broadcast information, it transmits a RACH burst in time slot 0 of the beacon frequency. The shortened burst of the RACH ensures that regardless of the distance from the cell, the burst will stay within the boundary of the time slot (remember, for normal bursts, timing advance is used to ensure that the burst is received in the right slot). This RACH contains identification data for the mobile and also allows the network to perform authentication to ensure the mobile has the right to be on the network.

From there, the mobile remains in the idle state, performing location updates as needed as it moves from cell to cell. Generally, as the mobile receives a broadcast from a new BTS, it knows it is in a new cell, and thus will go through the location update process. After being registered, the mobile can monitor the PPCH, looking for information intended for the specific mobile. Just as in voice, there is a sleep-mode operation. The mobile sleeps and wakes on a set schedule, generally determined by an algorithm based on its TMSI, such that it is monitoring the PPCH at set intervals.

If and when the mobile needs to establish a data channel, it enters the standby mode by sending a packet channel request message on the PRACH. The BTS then responds with a packet channel assignment message on the PAGCH. From there, the mobile can enter the ready or traffic state by transmitting and receiving the PTCH, in addition to the PACCH for signaling control. In addition, periodically, the mobile will transmit a RACH in order to allow the BTS to adjust the timing via the downlink PTCCH. Figure 6.13 shows the processes involved with downlink initated resource allocation.

6.12 The EDGE Format

EDGE was designed specifically as an upgrade to GPRS, for integration into GSM networks. In GPRS, the gross payload per time slot is 116 bits. In that same time slot

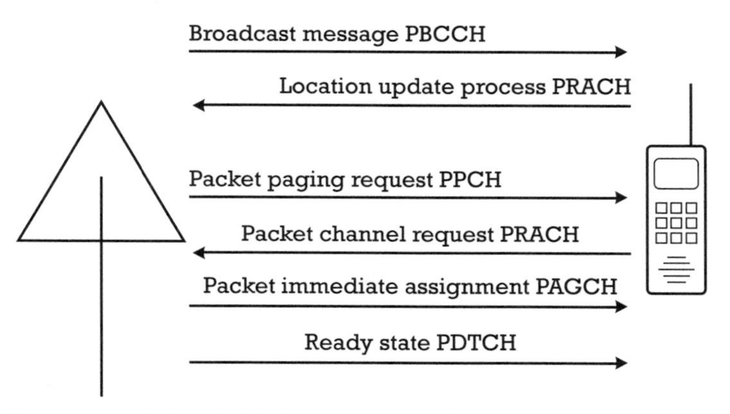

Figure 6.13 The process of downlink-initiated resource allocation.

with EDGE, the gross payload becomes 464 bits. This increase comes from the use of a higher order RF modulation format and different coding rates. GPRS used GMSK modulation, the same modulation as GSM. This meant that upgrading a GSM BTS to support GPRS was theoretically a software change, as it was the coding and software that was enabling the data rates. EDGE uses a new modulation format, *eight-phase shift keying* (8PSK), which means that upgrading to EDGE requires new hardware radio cards and is thus a bit more complex to install (see Figure 6.14). The packet technology, however, is identical in EDGE and GPRS.

6.13 8PSK Modulation and EDGE Coding

Previous chapters have spent quite a bit of time explaining the various modulation formats. By far, 8PSK is among the more complex of those discussed. In 8PPSK, each symbol change can represent 3 bits of information. In GSM and GPRS, each symbol shift would represent 2 bits, so this modulation allows for a higher rate.

The standard is designed such that the modulation can be interchangeable. For instance, one time slot can transmit data with GMSK, while the next time slot can transmit with the 8PSK. This eases the integration issues, as GSM-only phones will still be able to use EDGE-capable channels. Within the EDGE standard, either GMSK or 8PSK can be used for modulation.

One of the reasons higher order modulation formats are not used more is simply that they are not tolerant of poor RF environments. In other words, the mobile will need to have a higher RF received power in order to receive an 8PSK-modulated signal than it would to receive a GMSK-modulated signal. In addition, the signal can be affected more from fading due to movement. In fact, mathematically, the difference between the received levels needed for the same BER between GSM and EDGE is substantial enough that many system designers must consider more base stations for the same coverage and performance. Many operators are thinking ahead by

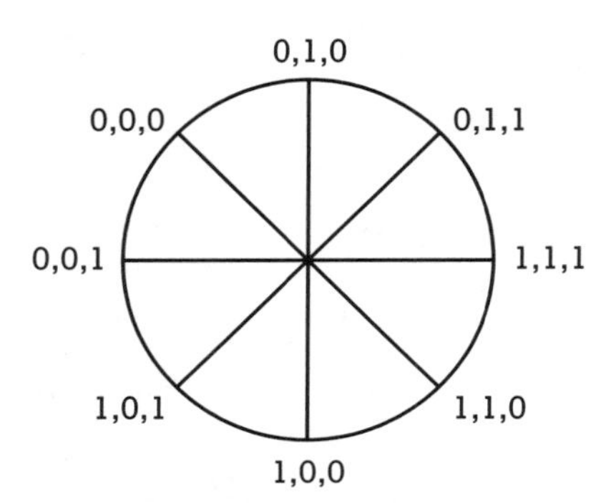

Figure 6.14 EDGE uses 8PSK modulation, where each phase change represents 3 bits.

deploying higher power amplifiers on the downlink and improving uplink sensitivity by using superconductor filters that improve BTS noise figure (and thus sensitivity).

Like GPRS, EDGE uses variable coding to allow for this, with the lowest data rates in EDGE matching the rates in GPRS. There are nine coding rates and modulation settings in EDGE, which allow for data transmission rates from 8.8 Kbps up to 59.2 Kbps per slot.

6.14 EDGE and the Future

Because EDGE is designed to transmit standard IP packets, any data that can be transmitted on the Internet can be transmitted fairly efficiently on the radio link. This leads to the concept of voice over IP, a concept many feel will not only become the norm for landline networks, but can also become the standard for wireless, using EDGE as the transmission medium. As the hardware will already be in place to get the bits from the network to the mobile, and vice versa, it is only a matter of using the right software and ensuring that the network can support an adequate data rate. Many operators have plans today to use EDGE for voice in the future.

At the time of this writing, most GSM operators have already deployed GPRS, and many have plans for EDGE deployment contingent on the need for higher-speed data. It seems obvious that if wireless data does become demanded, GSM operators will need to upgrade to EDGE to compete with CDMA2000's data capabilities and relatively easy higher speed data upgrades.

CHAPTER 7
CDMA IS-95 and J-Std-008

7.1 Introduction

CDMA-based formats are perhaps the most complex digital wireless systems in commercial use today. Rather than separating users with different frequency channels, as is the case with AMPS, or time slots, as with TDMA systems, CDMA puts all of the users on the same frequency at the same time, separating users with codes.

All of the primary 3G formats use a form of CDMA, making this chapter the foundation for the next generation (CDMA2000 and WCDMA) chapters.

CDMA2000 is really an upgrade to this format and is completely backwards compatible. For the purposes of this text, however, CDMA will be treated separately from CDMA2000, which will be covered in Chapter 8.

7.2 The Modulation We Use in CDMA

QPSK is the form of digital modulation used in CDMA. The QPSK carrier will cycle through four output phases, with all transitions being allowed, including through the origin. Like $\pi/4$ DQPSK, QPSK does not use a constant signal envelope, so amplitude will change as the phase shifts from one state to another.

Note that the symbol/phase correlation is not based on changes in phase (e.g., differential) but absolute phase position (i.e., each vector represents a specific symbol). Where $\pi/4$ DQPSK had four phase transitions, each representing 2 bits each, QPSK has four phase states, each again representing 2 bits each (see Figure 7.1).

7.3 Multiple Access in CDMA

As mentioned in Chapter 1, in CDMA the frequency remains the same for all users in all cells in the network, and all users transmit at the same time (except in the case where an extra frequency is added, but the concept remains the same). Users are identified by particular code and can transmit and receive at any time (e.g., there is no time domain, other than synchronization and the use of logical channels).

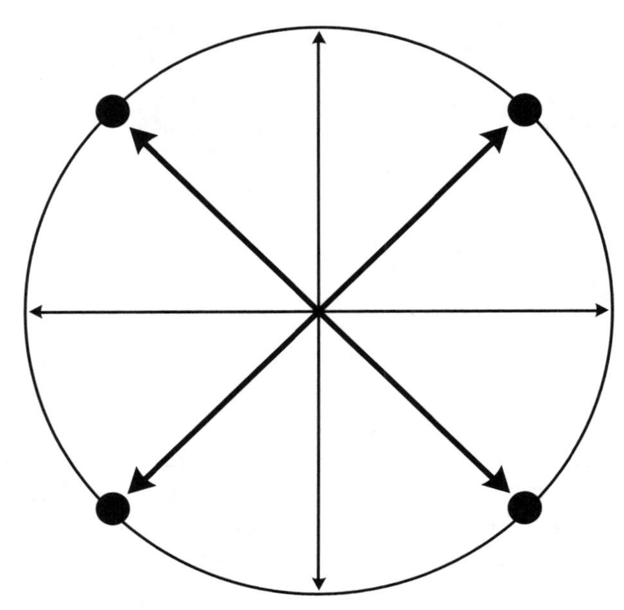

Figure 7.1 The black circles indicate the decision points in a QPSK signal. Each decision point represents 2 bits.

With frequency reuse not being an issue at all, adding cells to the network becomes substantially easier. This makes filling in holes or adding extra cells to cover special events a simpler task.

7.4 Capacity: The Reason for the Hype

No discussion on CDMA would be complete without spending a substantial amount of time on the primary design goal of the technology—increased capacity. Shannon's equation is the reason why spread spectrum systems are becoming so popular:

$$\text{Capacity} = \text{Bandwidth} \log_2 (1 + \text{Signal/Noise}) \tag{7.1}$$

What's important to see is that capacity is related to the S/N ratio and the bandwidth. Because we are using the same frequency band throughout the entire network, it is easy to see how problems with power in one sector can easily interfere with overall capacity of the system. Because the same frequency is used throughout the entire system, other sectors and base stations will sound like noise if a phone is not set to listen to them. Also, remember that because of processing gain, we can use a lower S/N ratio in spread-spectrum systems, while still maintaining the same quality of service.

7.5 Understanding the "Code" in CDMA

As mentioned earlier, CDMA does not use frequency or time to separate users. Instead, it uses different digital sequences, or codes, and in some cases different timings on the same sequence.

In CDMA, an often-used analogy is the "CDMA Cocktail Party." Imagine a lively party where everyone was speaking in a different language. Even with all of the people talking at the same time, you could still make out the person speaking your language, as long as there was only one person speaking it.

Now what are the catches to this concept? First, you would need to be located in the room where you would only clearly hear one person speaking your language. Also, the other languages being spoken nearby would need to be substantially different than your own [e.g., if you speak English, and someone nearby was speaking with a Scottish accent, and another near you was speaking with a New York accent, it would probably cause a problem as the languages (or codes) are not different enough from each other]. Also, the overall noise in the room needs to be low enough, in relation to the level you receive from the person speaking your language (the S/N ratio). If the language was different enough, and the person spoke intentionally very slowly and clearly, you might be able to get away with a lower S/N ratio (processing gain).

Thus, to get separation, we need to ensure the codes are different enough from each other. As discussed in Chapter 1, how well two digital sequences compare to each other is called the correlation. Remember, a correlation of one means the two sequences are completely alike. A correlation of zero means they are completely unlike—exactly what we need for separation of users. If two sequences have a correlation of zero, we call them orthogonal to each other. On the forward link, we separate users with 64 orthogonal codes, each being 64 bits long, called *Walsh functions* (see Figure 7.2 for how we derive these codes).

Hence, Walsh code 0 is all zeros. Each user will be assigned one of these codes, and this code will separate everyone (as well as the overhead channels). The next logical question is: How do we use these orthogonal codes to separate?

0

0	0
0	1

00	00
01	01
00	11
01	10

1. Repeat to the right and below.
2. Inverse diagonally.
3. Continue until 64 bits across and down.

Figure 7.2 Walsh functions are uncorrelated to each other. Note that Walsh code 0 (will be the top sequence, from left to right) will end up being all zeros when the pattern is taken out 64 bits.

The principle behind spreading and despreading is that when a symbol is XORed with a known pattern, and the result is then XORed with a known pattern, the original data will be recovered.

In cdmaOne, each symbol is XORed, or spread, with all 64 chips of the Walsh code. In Figure 7.3, the symbol of value one is spread with Walsh code 59, yielding a 64-chip representation of the symbol. In other words, for every one bit of data, we end up with 64 chips output. (Remember, spread spectrum was discussed in Chapter 1.)

In Figure 7.4, we spread each symbol with an orthogonal code of four bits for illustrative purposes. Thus, we output four chips for every one symbol of user data. We do this by XORing the user data with the Walsh code. Remember, in the real implementation, the 4-bit Walsh function shown is actually 64 bits long.

To recover the transmitted signal, we simply XOR with the correct Walsh code (the code used to originally spread), then we integrate the result.

If we were to use another Walsh code (which, of course, would be orthogonal to the actual code), you would see that it would not give us any usable data, and our system would immediately know that there is just noise on that channel.

The second type of codes used extensively in CDMA is PN codes, binary sequences that have the properties of randomness (i.e., equal numbers of zeros and

Information bit - "1"

0110011010011001100110010110011010011001011001100110011010011001
"Walsh code #59"

1001100101100110011001101001100101100110100110011001100101100110
The information bit "1" is now represented by this 64-chip sequence

Figure 7.3 In CDMA, each information symbol will be spread with all 64 bits of the Walsh code sequence.

1	0	0	1	1	User data
0110	0110	0110	0110	0110	4-bit Walsh function
1001	*0110*	*0110*	*1001*	*1001*	Result of spreading—transmit/receive chips
0110	0110	0110	0110	0110	4-bit Walsh function
1111	0000	0000	1111	1111	Result of despreading
1	*0*	*0*	*1*	*1*	Integrate—user data recovered!

Figure 7.4 The process of taking user data, spreading it with a sequence (XORing), transmitting it, receiving it, and then despreading (XORing) it with the same sequence.

ones, set number of runs). We use three PN codes in CDMA: two *short codes* and one *long code*. Before we see how these codes are used, it is important to understand some of the traits of PN codes.

A very important feature of PN codes is that if we time shift the same versions of a PN code, we end up with two codes that are close to completely uncorrelated to each other (nearly orthogonal). Thus, we can use the same sequence and time shift the start of the sequence, ending up with nearly orthogonal sequences (see Figure 7.5). The sequences that we use are all synchronized to specific triggers, which are derived from the GPS satellite system.

In order to offset a PN code to create nearly uncorrelated sequences, we use a masking system. The pattern will be the same, but the timing will be different. With different time offsets, each sequence will be nearly orthogonal to each other. Remember, the sequence remains the same, but it simply starts at a different point in the sequence. In the circuit shown, and sequence of bits placed in the "mask registers" will offset the output sequence a specific number of bits. The short PN sequence is 32,768 bits long, and we use increments of 64 bits to offset. This leaves us with 512 unique time offsets on the sequence. Also, remember that because the short code is always transmitted at the CDMA data rate (1.2288 Mbps), the sequence always take 26.667 ms to completely cycle, regardless of the time offset. As stated earlier, the start of the sequence is triggered by a unique trigger from GPS called the *even second clock*.

7.6 The CDMA Vocoders

As discussed in Chapter 1, a vocoder is a device that can take analog voice and, using various predictive algorithms, compress and encode this voice data.

Currently, three vocoders are supported in CDMA systems. Originally, the 8-K vocoder was to be the main vocoder. Remember, a lower data rate means better system performance (but can sacrifice voice quality) because a lower data rate gives the system more processing gain. Carriers decided the degradation in voice quality was

0 0 1 0 1 0 1 1 0 0 1 1 0 1 1 Sequence A

1 0 1 1 0 0 1 1 0 1 1 **0 0 1 0** Sequence A with a 4-chip offset

Both sequences synchronized to a standard clock reference.

Figure 7.5 The same sequence with a time offset. These two sequences are now nearly uncorrelated.

not worth the gain in capacity, and thus many carriers initially used the 13-K vocoder.

The 13-K vocoder had excellent voice quality; however, it significantly lowered the performance of the network in terms of capacity. The *enhanced variable-rate vocoder* (EVRC) is an enhanced 8-K vocoder—the enhancement is in the form of improved voice quality. If the effective bit rate goes down while the transmitted data rate remains the same, the processing gain goes up; the S/N ratio can then be lowered for the same capacity or kept the same to allow for more capacity.

CDMA uses a variable vocoder. If you remember that the effective data rate affects our E_b/N_0, we can get a significant gain by dropping our data rate from the full rate of 13 or 8 Kbps to half-rates or even eighth rates.

In a normal voice conversation, one person speaks while one person listens. Even when we are speaking, there are pauses between our words. Variable vocoders take advantage of this fact by varying the data rate, so that high data rates are only needed when they are necessary—during actual voice periods.

Both data sets produce a frame every 20 ms (50 fps) using code excited linear prediction. Rates are either full, half, quarter, or eighth. This means that every 20 ms, the system makes a determination on what data rate to use. This determination is also tied into a power bursting system (which will be described later in the chapter), so that not only is the data rate lowered, but the actual output power is lowered as well during lulls in speech activity. Remember, if we can lower the data rate, we increase our processing gain, and if we lower our output power, we lower the overall noise in the system—both are keys to improving capacity.

7.7 The Forward Link—Code Channels

There are 64 code channels in each 1.2288-MHz CDMA channel (see Figure 7.6). Of these 64, at least three must be overhead channels:

- Pilot channel—always Walsh code 0;
- Paging channels—always Walsh code 1, can be up to Walsh code 7, depending on system requirements;
- Sync channel—always Walsh code 32.

After a call is up, each user is assigned one forward traffic channel. These account for the remaining codes.

Figure 7.7 shows the final stage of the forward link. It is the same for all of the forward link code channels. You will notice that already-spread data is sent to two channels: the I and Q. These two channels then spread the data again using the short PN codes (remember, these short PN codes have a period of 26.67 ms), which are offset (in 64-chip increments—512 possibilities) depending on the sector from

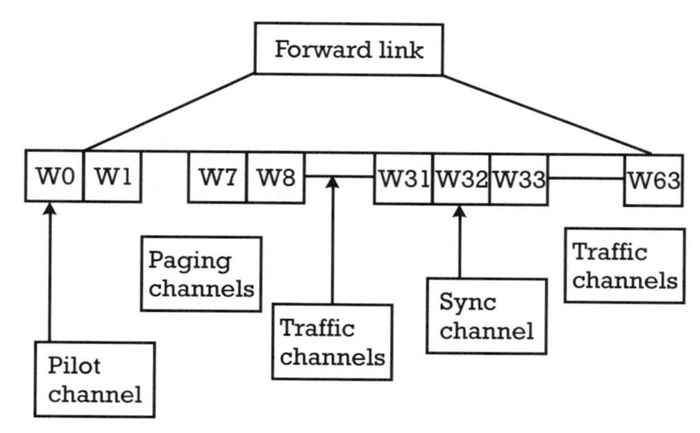

Figure 7.6 The forward link, all in the same 1.2288-MHz channel, consists of 64 Walsh-coded channels.

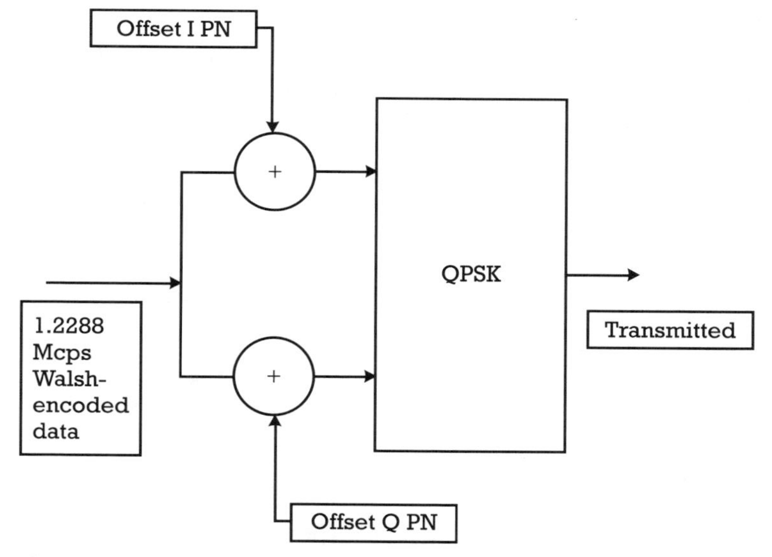

Figure 7.7 The final stage of the forward link channels, showing the short PN code spreading. This offset separates base stations/sectors.

which they will be transmitted. This top layer of spreading acts as an identifier. Each sector is assigned its own offset, similar to SAT in AMPS.

The pilot channel (see Figure 7.8) acts as a timing beacon. It has no information on it. This figure is used to show that the Walsh function zero's property of being all zeros is significant. Because the data is all zeros, the Walsh function is all zeros; thus, the result of this spreading will be all zeros. If you spread a sequence with all zeros, it simply passes the sequence through. This means the pilot channel transmitted data is simply always the short PN codes, with the specific PN offset. It is used by all

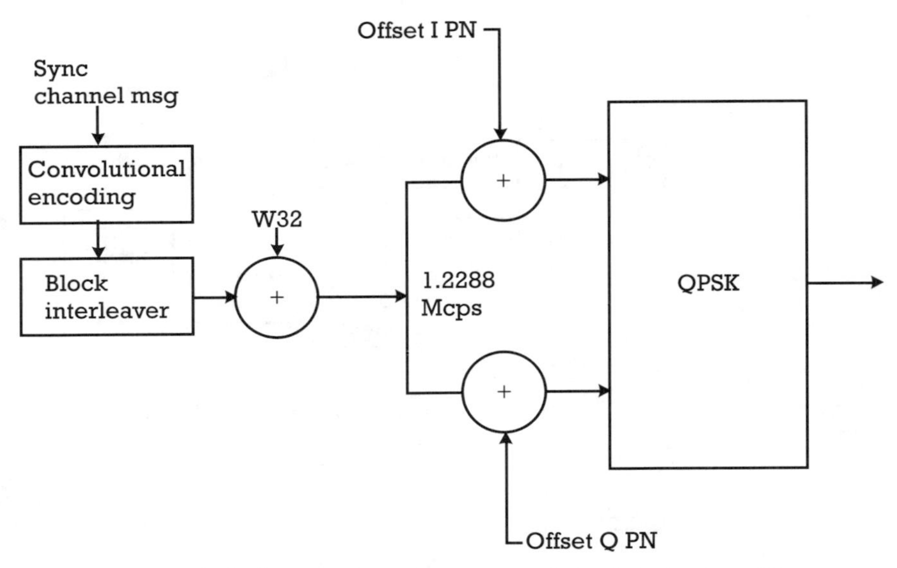

Figure 7.8 The basic block diagram of the pilot channel. It simply is the short PN sequences with the unique time offset particular to the base station/sector from which it is transmitted.

mobiles as a coherent phase reference, as well as a means to identify each cell/sector and compare strengths for the handoff process.

Because the pilot sequence is just the short code sequence (offset with whatever PN offset designated for that sector), it repeats every 26.67 ms and is triggered by *global positioning system* (GPS) timing references. This allows mobile stations to easily synchronize to it, as they know what the sequence is (the short PN code), just not the offset. The pilot contains about 20% of radiated power.

The sync channel is used to synchronize with the system. It always uses Walsh 32. The sync channel is comprised of an 80-ms superframe, divided into three 26.667-ms frames. These frames correspond to the length of the pilot sequence and use the same triggering (the GPS timing references). It transmits the following information: system time, pilot PN of base station, long-code state 320 ms into the future, system ID, network ID. Of the three overhead channels (pilot, paging, sync), the sync has the least amount of power in it.

Knowing that the pilot sequence will take exactly 26.667 ms, that the sync channel frame is also 26.667 ms, and that they are both triggered by the same reference (the GPS timing signals), it is easy to understand how a handset can first acquire the pilot channel, compute the start and stop of the pilot sequence, and then switch to Walsh code 32 to decode the sync channel information. Thus, they obtain the absolute timing information needed in order to begin coherently demodulating the paging and traffic channels on the forward link (as well as establish timing for its own transmissions).

Information on the sync channel includes:

- Protocol versions;
- SID;
- NID;
- PN offset;
- Long-code state;
- System time;
- Local time;
- Paging channel bit rate (either full or half).

The paging channel (Figure 7.9) is the DCCH of the forward link. It uses Walsh codes 1–7 (always at least 1), and it used to page the mobiles, transmit overhead information, and assign mobiles to traffic channels. Information on the paging channel is either sent at full rate or half-rate.

Similar to NA-TDMA, CDMA has a sleep mode while the mobile is monitoring the paging channel. The network will place the messages destined for a particular mobile station in a specific time slot in the paging channel. The paging channel is divided into 2,048 slots, each 80 ms long, for a total cycle time of 163.84 seconds. Sleep mode is established by allowing the mobile station to wake up during specific

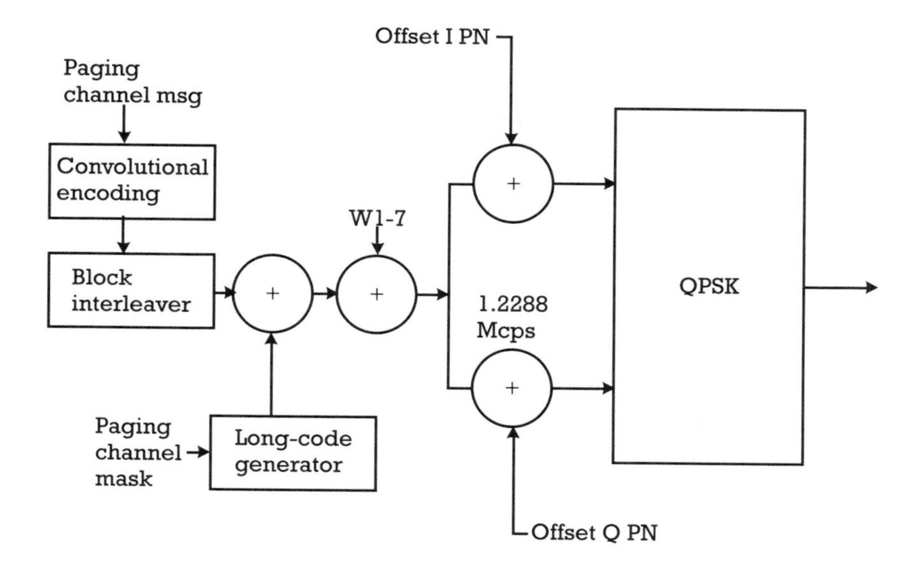

Figure 7.9 The basic block diagram of the paging channel.

slots. Which slots in the channel the mobile station will monitor is determined by a mathematical function based on the MIN and ESN of the mobile station.

In addition to the standard messages similar to those seen in AMPS and NA-TDMA, such as the PAGE, RELEASE, authentication, and registration messages, and maintenance messages, the following messages can also be sent and are somewhat unique to CDMA:

- *Systems parameter message:* This includes various setup information (e.g., which type of registrations are in effect).
- *Neighbor list:* This contains the PN offsets of nearby sectors.
- *Access parameter message:* This is similar to the global action message in AMPS; it sets up access protocols, which access channels to use.
- *CDMA channel list message:* This tells the mobile station whether any other CDMA carriers are in the system.

Generally, every one paging channel can handle up to 32 traffic channels. After synchronizing to the BTS, the mobile monitors this channel in the idle state.

You will notice in Figure 7.9 that before the paging channel message is spread with its Walsh code, it is spread with the long code with an offset mask that includes the paging channel number and PN offset of the sector from which it is being transmitted. This added level of security is also used in a similar manner on the traffic channels.

The forward traffic channel (see Figure 7.10) is used to transmit traffic (voice) data as well as signaling information during a call. Note that the long code is used with the user-specific mask to identify each individual user. The Walsh code can be any of the remaining codes (up to 55). Once this traffic channel is assigned to a user in an omnicell or sector, it cannot be assigned to another user for the duration of the call.

Like the blank-and-burst and FACCH processes in AMPS and NA-TDMA, CDMA interrupts voice data during a call to insert signaling messages. There are a number of ways these messages can be sent. The blank-and-burst process in CDMA, like in AMPS, substitutes signaling for all of the voice. For less important information, the dim-and-burst mode can be used. In this mode, only some of the user information is overwritten. During dim-and-burst and black-and-burst processes, only the full-rate data rate is used.

Messages sent on the traffic channels are very similar to those in NA-TDMA, including authentication messages (remember, the authentication process in CDMA is the same as NA-TDMA), flash and alert messages, and various similar call-management messages. A few messages are added, mainly to support the complex handoff process and added security measures. (CDMA allows for transitions of the long code in order to heighten security.) There are also messages related to what are

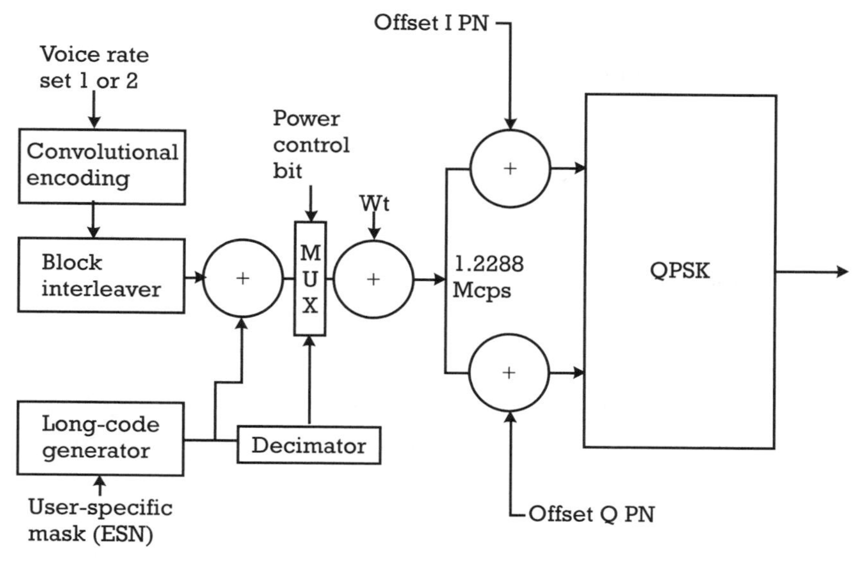

Figure 7.10 The basic block diagram of the forward traffic channel.

called *service options*. These pertain to different service setups using the two vocoders (e.g., a test set can send a message to a handset instructing it to go into a loop-back mode with rate set two).

Built into the forward traffic channel is the power control subchannel. This subchannel is actually one bit per 1.25 ms, or 800 times per second. This bit is punctured over the data, after interleaving and encoding, in order for the mobile station to be able to receive the bit quickly. The power control bit actually replaces two of the information bits, with the effect that the power control bit will have the energy of 2 bits. The mobile responds to the power control bit by increasing or decreasing its transmit power by 1 dB.

You may also realize that the sync, paging, and traffic channels do not always transmit at full rates, yet the output data rate is always 1.2288 Mcps. This is done by a process of repetition, as well as in the previously mentioned process of putting additional messages into the frame.

All of these various forward link channels are summed together into the same 1.2288-MHz channel and sent out in the same sector/omnicell with the same PN offset.

7.8 Synchronization on the Forward Link

First, the mobile finds a pilot channel and syncs to the start and finish of the PN (exactly 26.67 ms). Next, the mobile is able to read the sync channel message (in which each frame is exactly 26.67 ms) and fully synchronize to the system. The mobile can then monitor the paging channel and get system and access parameters. If a call comes up, the traffic channel goes active and the mobile station moves to decode the appropriate Walsh channel.

The signal is first downconverted, and A/D is performed to get the signal to digital baseband. The Rake receiver (see Figure 7.11) is then used, coherently combining the multipaths to create one more robust digital baseband signal. The signal is then descrambled, deinterleaved, and Viterbi decoded. It actually decodes the vocoded data at all four rates and then uses metrics to decide which rate was the most likely rate transmitted.

The Rake receiver is a unique part of the CDMA system that allows a receiver to search for strong multipaths as well as other base stations and slide in time to find their code. Once found, the receiver element can demodulate the signal, coherently combine the signal to create a more robust receiver, and allow the radio to "make before break" on handoffs. Most mobiles have three receiver elements (or fingers). Most base stations have four fingers.

7.9 The Reverse Link Channels

Two types of channels are in the reverse link (generated in the mobile station): the access channel and the traffic channel. In the reverse link, handsets are separated by unique offsets of the long code (remember, in the forward link, base stations were identified by unique offsets of the short codes on the I and Q channels, and channels/users were separated by Walsh codes).

The reverse channel (see Figure 7.12) uses a zero offset on the I and Q PN short codes. It also uses a half-chip delay on the Q channel. This half-chip delay allows for easier amplifier design, as it ensures there are no transitions through the origin during phase shifts. This is the reason the reverse channel modulation is called offset QPSK. If you can picture the I and Q channels as the X and Y portions of a graph, if

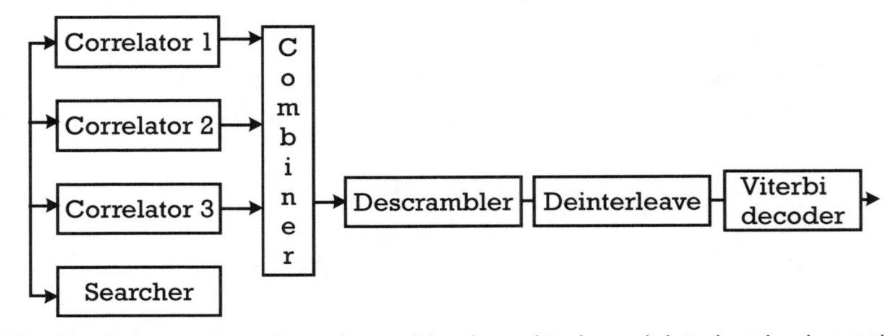

Figure 7.11 The Rake receiver allows for multipaths to be demodulated and coherently combined for a more robust signal, where normally destructive fading would occur.

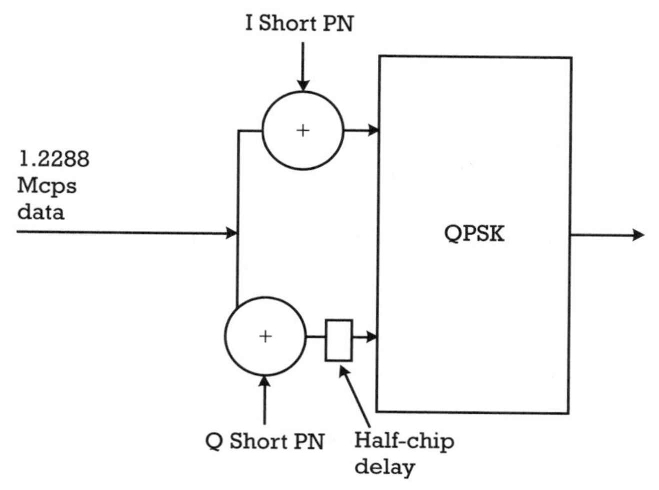

Figure 7.12 The final stage of the reverse channel process. The half-chip delay ensures transitions do not pass through the origin.

you move from one vector to another you might normally get there by changing the X and Y position simultaneously. In this offset manner, we move the X first, then the Y, so the movement is not a direct line.

The reverse channel is significantly different than the forward channel (see Figure 7.13). An orthogonal modulation scheme is used, not for user separation, but to spread the signal and allow for easier noncoherent symbol detection. Users are

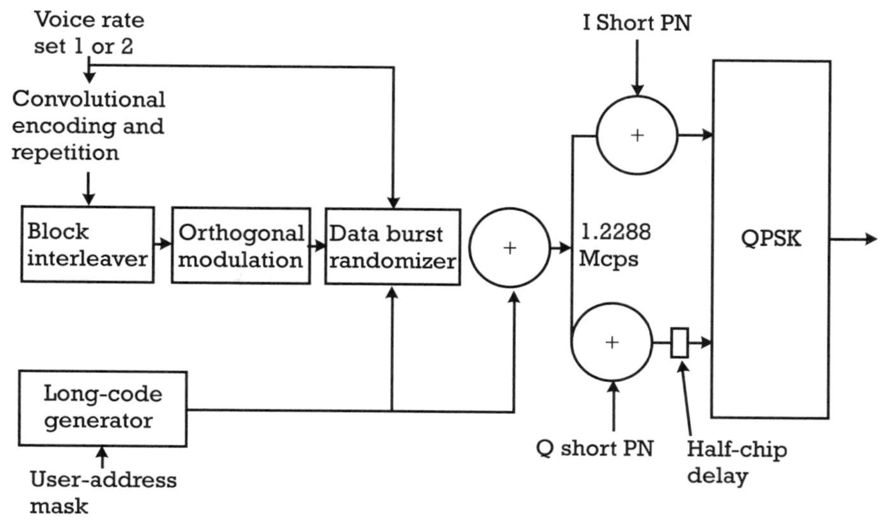

Figure 7.13 The basic block diagram of the reverse traffic channel.

separated via a mask based on the ESN of the handset on the long code, which spreads the data rate to 1.2288 Mcps. This long code adds a serious degree of security protection in addition to separating the users. Repetition is used during lower data rates in order to get the data rate constant for transmission.

The mobile station bursts its power to save power and also lower the overall noise in the network. If a mobile does not have to transmit power continuously, it requires less battery power, and thus talk times are extended. Also, because the mobile station's transmitted energy will act as noise to other mobiles in the cell, it is important to keep the output power as low as possible. Bursting the power accomplishes this as well. The mobile station makes decisions on what data rate to use every 20 ms, and this decision is passed on to the data-burst randomizer. The bursting is randomized to spread the transmitted power over time. Also, the transmitted power is lowered 3 dB (halved) for each lower data rate. At full rate, there is no bursting. Note that the data is repeated at these lower data rates, so no data is actually lost by bursting the power. The signal is bursted in 1.25-ms *power control groups* (PCGs), and there are 16 PCGs in every 20-ms frame (see Figure 7.14).

As mentioned earlier, there are two types of reverse channels: the traffic channel, used during a conversation, and the access channel. The access channel is used to originate calls, for registration, and to respond or acknowledge paging channel messages. The access channel is differentiated the same way the reverse traffic channels are, via the unique mask on the long code. This mask includes a specific channel number given to the mobile by the base station. The mobile station sends the access channel message to the base station, called an access probe, and waits for an acknowledgment on the paging channel. If it does not receive the acknowledgment, it waits a random amount of time, increases its power, and tries again. The number

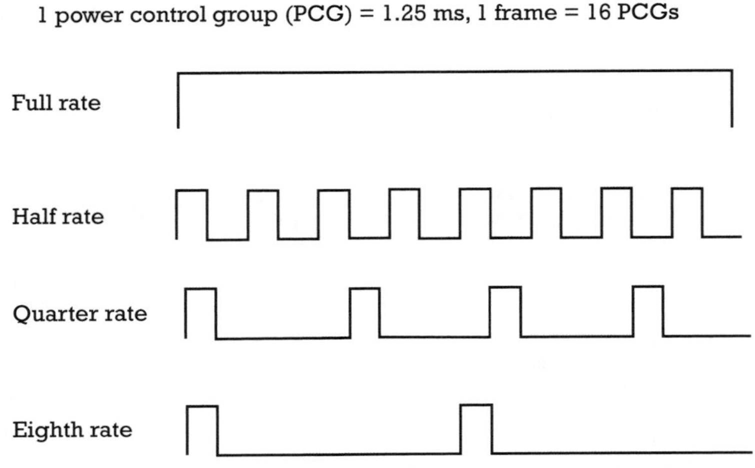

Figure 7.14 The mobile station bursts its power in 1.25-ms PCGs. Decisions are made for every 20-ms frame depending on voice activity.

of probes to try in each *sequence*, the number of sequences to try per *attempt*, and the amount of power to increase after each unsuccessful probe are all set by the network via the ACCESS PARAMETERS message (see Figure 7.15).

7.10 Power Control

Physically, the forward link is significantly different than the reverse link. First, all of the code channels take the same path to the mobile, so they all fade together. Thus, at the handset receiver, all of the channels are arriving at the same power levels and do not fade independently. The forward link uses better codes (Walsh codes) to separate users. This ideal, perfectly orthogonal separation makes it much easier to pull out your channel's data. Using the pilot and sync signals, the mobile is able to coherently demodulate using nearly perfect timing and phase information. Thus, a dynamic range of just 4 dB is sufficient.

In the reverse link, the characteristics are very different. Mobiles closer to the base station may be up to 80 dB higher than a mobile further away, drowning out the far away handset. Also, each mobile will have a different path, a different loss, and a different fade. Also, as there is no pilot, the reverse channel must be demodulated noncoherently.

Thus, the reverse link power control is much more extensive. The goal is to have every mobile reach the base station at the same level, which requires a very complex system.

The initial power control system is the open loop. The open loop gives the mobile a starting point for beginning transmissions. This gives the closed-loop process a head start, allowing the mobile to get to its ideal power much quicker.

Essentially, we take the difference between the average path loss (–73 dB for the cellular bands and –76 dB for the PCS bands) and mean received power (which is the

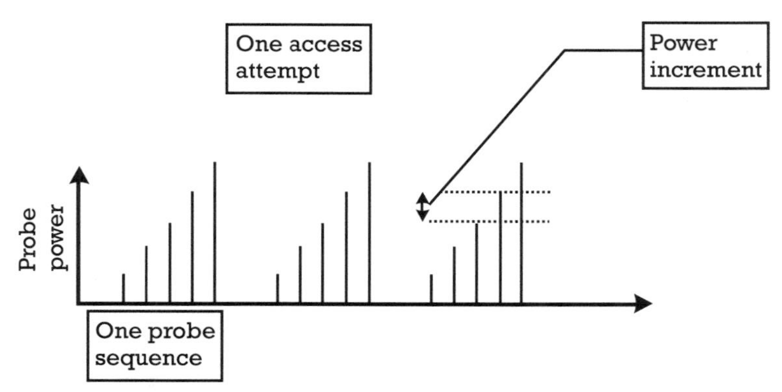

Figure 7.15 The mobile station sends access probes in a manner designated in the ACCESS PARAMETERS message.

power needed to compensate for path loss) and add in any variants given by the base station. This is then the starting level. It will provide accuracy within ±6 dB.

The closed-loop power control involved the use of the power control bit punctured into the forward traffic channel. Because the reverse traffic channel uses 1.25-ms PCGs, the object is to adjust the power for each PCG. This is done by measuring the received power at the base station and then puncturing a 1-bit command (either up or down) every 1.25 ms, or 800 times per second. The exact timing of the *puncture* is randomized by using certain digits of the long code.

Finally, CDMA uses an *outer loop*. The outer loop works with the closed loop. In the closed loop, the base station needs to react very quickly (800 times a second is pretty quick). For this reason, the base station does not have time to send the PCG based on *frame-error rate* (FER)—the true determinant of call quality. Instead, it simply measures received power, compares it to a threshold, and then determines whether to increase or decrease the mobile's power. It is the job of the outer loop to monitor actual FER and adjust the threshold accordingly. The outer loop also serves as a trade-off manager. That is, in CDMA, operators can trade off some quality in order to increase their capacity. As more users establish calls on a base station, for instance, the outer loop can lower the threshold, which may in fact increase FER on the calls slightly but allow more calls onto the system. Generally, the outer loop has some settable parameters by the operator, mainly a lower and upper limit, as well as a midpoint. In this way, an operator can control how much trade-off occurs.

7.11 Call Processing—The Four States

Figure 7.16 shows the basic call-processing loop. After power up, the initialization state determines which system to use (whether analog or CDMA). If it is CDMA, it goes into pilot and sync processing. Once the system is synchronized, the system

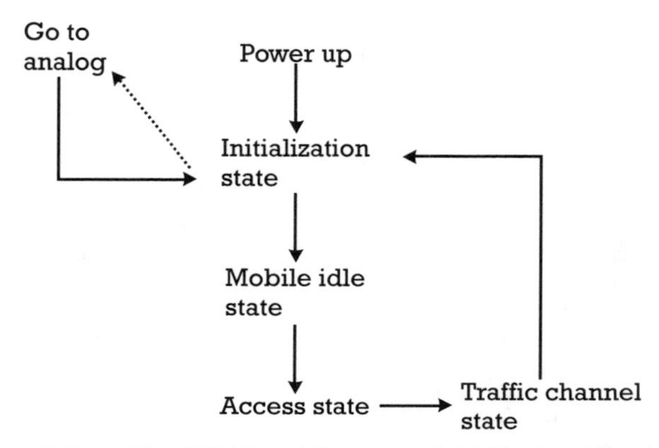

Figure 7.16 The four states of the CDMA mobile station: initialization, idle, access, and traffic channel states.

goes into the mobile station idle state, where it monitors the paging channel. If a call is to be originated or the mobile is paged, the system goes into the access state.

Once a call is set up, the phone moves over to the traffic channel state, where the forward and reverse traffic channels are used to communicate voice and messaging.

In the initialization state, after determining that there is a digital system, the handset monitors the paging channel. In determining the start and finish of the pilot channel, it can determine the timing of the sync channel. Once it can read the sync timing, it can further refine its timing (see Figure 7.17).

During the idle state, the mobile will monitor the paging channel. Various messages pertaining to set up and operation are on the paging channel.

Certain situations will trigger the mobile to drop out of the traffic state (drop the call on purpose):

- *Mobile ACK failure:* Certain messages require an ACK; generally, a mobile will retransmit the message after 400 ms, but if no ACK comes after three tries, the mobile drops the call.
- *Base station ACK failure:* This is similar to the mobile ACK failure, but it is not standardized.
- *Mobile fade timer:* The timer is set to five seconds after receiving two consecutive good frames. If the timer gets to zero, the call drops.
- *Mobile bad frames:* If there are 12 consecutive bad frames, the mobile drops the call.
- *Base station bad frames:* This is similar to mobile bad frames, but not standardized (i.e., manufacturers can implement this however they choose).

7.12 Registrations in CDMA

As discussed in previous chapters, registration is the process in which a mobile station makes its whereabouts known to the system. Without registration, a mobile

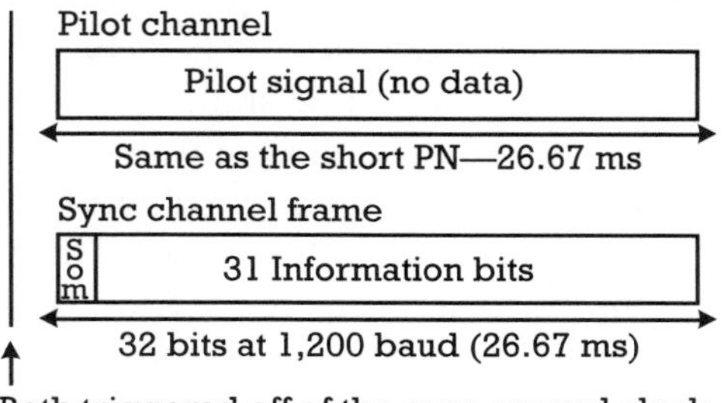

Figure 7.17 By synchronizing with the pilot sequence, the mobile can easily synchronize enough to read the sync channel.

would have to be paged across the entire system. On the other hand, having a mobile register every time it crosses into a different sector would load up the access and paging channels too much.

CDMA offers a variety of registrations methods that can be set by individual carriers, independently:

1. Autonomous:
   - Power up—whenever the phone turns on;
   - Power down—before the mobile turns off;
   - Timer based—registers when a timer expires;
   - Distance based—registers when the distance between the current cell and the last cell registered exceeds a current threshold;
   - Zone based—whenever a new zone is entered.
2. Nonautonomous:
   - Ordered—a base station that becomes aware of a mobile station for which it does not have all of the information on it can order the mobile to register;
   - Traffic channel—registration while on a call;
   - Parameter change—if certain parameters change, the mobile will register;
   - Implicit registration—occurs when a base station and mobile station exchange messages not directly related to registration, but convey enough information for registration to take place (e.g., an origination or page response message).

7.13 Handoffs in CDMA

This section will discuss handoffs. Handoffs in CDMA are substantially different than in other formats. Additionally, this section will discuss the parameters for the handoff function, which is quite configurable from a carrier's standpoint and is very important in the network optimization process.

Soft handoffs occur when a mobile makes a link with a new base station before it breaks the link with the current base station. In CDMA, the mobile will continuously look for alternate PN code offsets in order to detect possible handoff candidates. The mobile can than request a soft handoff (depending on the S/N ratio of the pilot), and the BSC can direct the soft handoff. The mobile will receive both signals and coherently combine them. The base station will receive vocoded frames from both base stations and determine which is most error free to use.

A softer handoff involves two sectors of the same base station. Generally, sectors will share hardware from the Walsh encoding, so one channel element can support multiple sectors (see Figure 7.18).

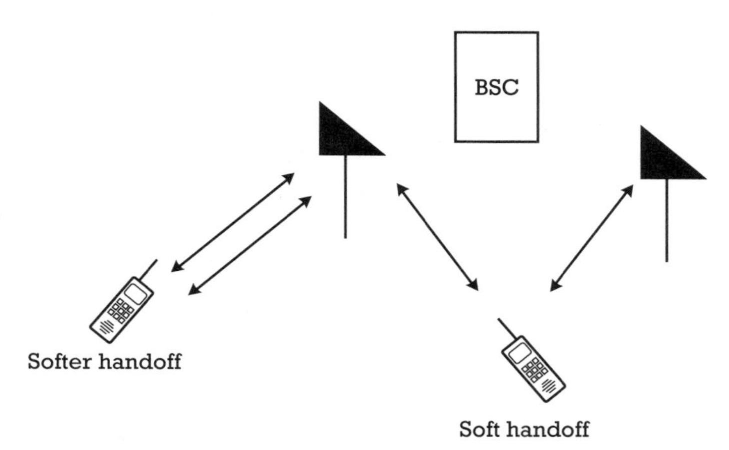

Figure 7.18 In a soft handoff, it is make before break, as opposed to the hard handoff—break before make.

A hard handoff involves a complete disconnection from the serving cell before connecting to the handoff cell. Whenever a mobile crosses from a CDMA network into an FDMA network (IS-95 is a dual standard), a hard handoff must occur. Also, some manufacturers do not support inter-BSC soft handoffs. Hard handoffs may occur because of frequency differences within or out of the network (e.g., a wireless operator may add 1.2288 channels to the system to increase capacity).

Hard handoffs typically require much more power to be transmitted; hence, they often wreak havoc with capacity and are not wanted in a network. When they do occur, it is essential that they work perfectly, as this is a cause of many dropped calls.

During the idle state, the mobile will perform soft, softer, and hard handoffs as needed. These are known as idle handoffs.

7.14 The Pilot Searching Process

The mobile keeps track of pilots in a special database. The pilots are grouped in the following sets:

- *Active set:* This includes pilots associated with the forward channel currently assigned to the subscriber station.
- *Candidate set:* This includes pilots whose strength exceeds a threshold but have not been assigned.
- *Neighbor set:* This includes pilots transmitting in the vicinity. It is given to the mobile by the BSC.

- *Remaining:* All of the pilots possible in the current system.

The propagation delay between the base station and the mobile is not known. Also, the multipath effect will cause varying amounts of delay between multipaths of the same signal. As a result, the mobile receiver must shift its PN code generators to search for multipath arrivals. Search windows define how much the receiver should vary the PN code in PN chips. It is a parameter set by the BSC.

There are three specific search windows for the different sets of pilots, and the search windows range from four chips to 452 chips. Obviously, there is a trade-off between increasing the search speed and perhaps missing a strong multipath that is outside the window. Setting up the search window parameters correctly is part of the optimization process.

Figure 7.19 is a graph that shows two cells as viewed by a mobile. Note some of the parameters, which are all configurable by the network:

- *T_ADD:* This is the threshold at which a pilot is added to the candidate list.
- *T_DROP:* This is the threshold at which a pilot is dropped from the candidate list.
- *T_COMP:* This is the comparison threshold. When the strength of a new pilot exceeds it, the mobile notifies the base station (not shown).
- *T_TDROP:* This is a time threshold that must tick off before a pilot is dropped from the candidate/active set.

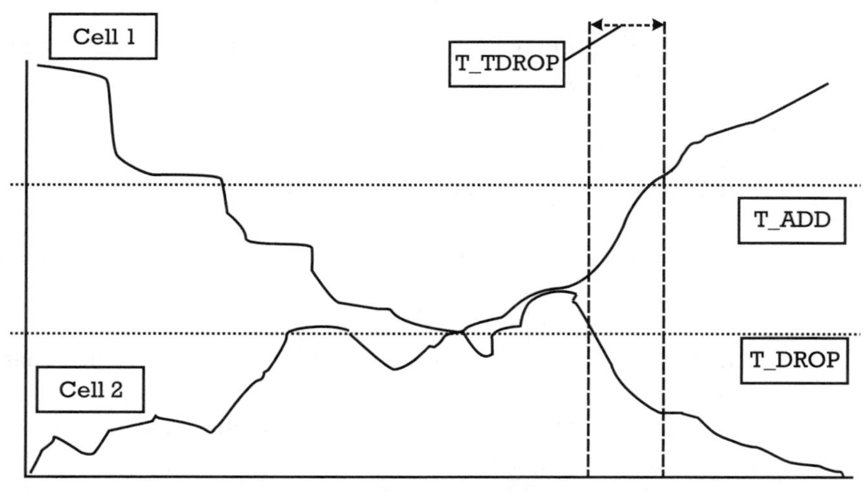

Figure 7.19 A graph showing two pilot measurements by a mobile station. The mobile is moving toward cell 2, away from cell 1.

7.15 Conclusion

CDMA is one of the most complex formats in commercial use. This chapter was designed to give a basic technical overview of the first generation of this format in order to facilitate the understanding of how to test CDMA systems. CDMA2000, the next generation of CDMA, is a backwards-compatible upgrade and will be discussed in Chapter 8. WCDMA, while not backwards compatible with CDMA, uses many of the same concepts and features.

CHAPTER 8

CDMA2000

8.1 Introduction

In the early to mid-1990s, most cellular industry professionals around the world found themselves taking sides in the CDMA versus GSM argument. Primarily, GSM advocates touted universal acceptance of the standard, which allowed for easy roaming—essential in the mobile European environment—as well as economy of scale, which made GSM infrastructure and mobiles less costly. CDMA advocates generally mounted their case on the capacity benefits of their technology in addition to the ease of upgrading to next generation technologies—especially higher-speed data.

In hindsight, it is hard to say whether there was a clear-cut winner or loser. While GSM continues to dominate on a worldwide scale, CDMA made inroads—probably more than most people expected—including deployments in Japan, Korea, China, Australia, Canada, Eastern Europe, across Latin America, and the United States.

Today, the importance of the original arguments has certainly shifted. Wireless operators today must focus almost all of their attention on capacity, as many are running out of spectrum to add more channels. In addition, the long-awaited and much-touted need for high-speed wireless data may indeed be at hand. It would seem that many of the growing pains those wireless operators who chose to use CDMA rather than GSM went through might have been worth it. The reason ties into the CDMA standard's ease of upgrading, allowing for a relatively inexpensive and technically simple upgrade to next generation versions—versions that allow for substantial capacity gains as well as higher-speed data, while at the same time maintaining complete backward compatibility.

8.2 CDMA Evolution

CDMA2000 is the term generally used to describe the next generation versions of the IS-95 standard. Many industry types might say it refers to the entire upgrade path of CDMA, which has been developed to be deployed in stages as the technology becomes available and the demand from the market necessitates it. Looking at the original IS-95 standard to the complete 3G version, there are six primary steps a CDMA operator might consider:

1. IS-95: The standard 2G format heavily deployed.
2. IS-95B: This has been deployed in Asia but not widely deployed in the United States. Incorporates several enhancements as well as packet-switched data.
3. 3G-1X, also known as radio transmission technology with one carrier (1XRTT), or the first phase: This is often just called 1X. It is the first 3G phase of CDMA, expected to be widely deployed initially because of capacity gains and for its high-speed data capabilities. Many are calling this a 2.5G standard.
4. 1XEV-DO, also known as Qualcomm *high data rate* (HDR): This is a high-speed data-only solution that can be integrated relatively easily into existing CDMA networks. Again, many consider this a 2.5G interim solution. There is a thought that 1XEV would be deployed in parallel with 1X, thus using 1X for voice services and 1XEV for data services.
5. 1Xtreme or 1XEV-DV: An enhanced version of 1XEV-DO, it also allows for simultaneous voice services.
6. 3G-3X (3XRTT): The complete 3G version of CDMA. Currently expectations are that it will only see limited deployment because of high deployment costs and demand and competitive issues.

The key to the six versions is the integration aspect. Wireless operators should be able to deploy these versions relatively easily, with relatively low investment. Note that this is qualified with "relatively." Upgrading is not a free process, nor is it completely plug and play, technically. Compared to the completely new deployment of a brand-new technology, such as many GSM operators face with deploying WCDMA, the cost and complexity should be lower.

IMT-2000 is a global standard designed to harmonize the third generation standards. One of the primary goals of this standard is to standardize the type of backhaul network on which the radio formats can operate. You will remember from earlier chapters that IS-95 is designed to operate on the IS-41 network, while GSM networks work on the MAP networks. As CDMA moves forward in evolution, it will be possible to operate it on either GSM-MAP or IS-41 networks (see Figure 8.1).

8.3 Overview of 1X-RTT and 3X-RTT

Almost every CDMA operator in the world today is planning to deploy at least the first advanced phase of CDMA2000, known as 1X. The 1X implementation should

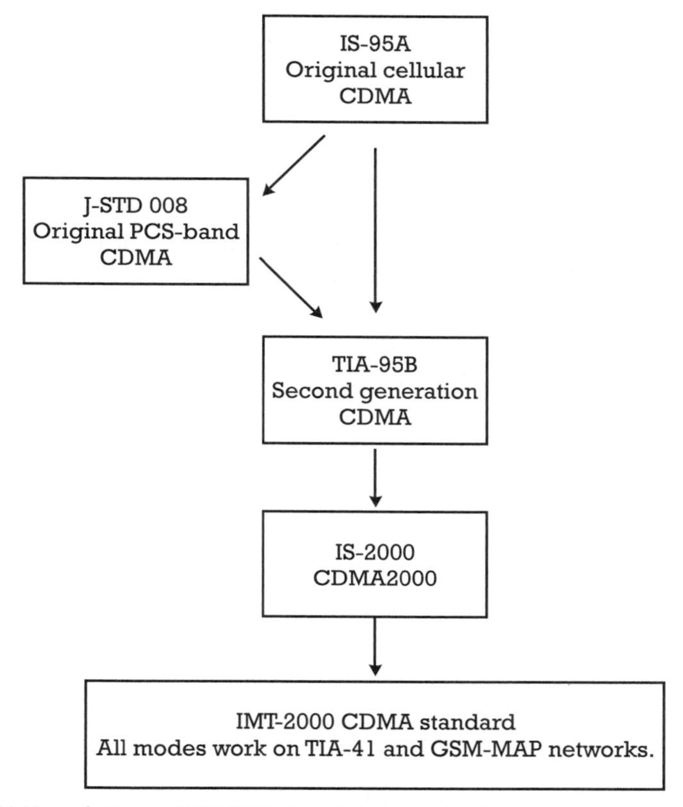

Figure 8.1 CDMA evolution to IMT-2000, the global standard.

be relatively simple. Because the bandwidth of the signal remains identical (1.2288 MHz), and the system is completely backwards compatible, carriers can deploy the technology—in most instances as a circuit card and software upgrade to existing base stations—and realize the benefits of the deployment as they swap out the older handsets for 1X-compatible handsets.

As mentioned, the benefits of deploying 1X infrastructure can only be realized if the end users are also given 1X-capable handsets; otherwise, the network will simply perform as an IS-95 network.

Given the ease of deployment, the reasons for deploying 1X-RTT are quite compelling. Under IS-95A, operators could deploy voice services, utilizing up to 64 Walsh codes. In addition, the capability is available for circuit-switched data, albeit at very low speeds. IS-95B offered some improvements. While it still only allowed 64 Walsh codes, it allowed for packet-switched data by combining Walsh codes for higher data rates. In this way, data rates of 64 Kbps or 96 Kbps could be deployed and offered. The obvious problem with this is the limit of Walsh codes—if several

Walsh codes were used for data, this would have an adverse effect on capacity of that carrier.

The 1X technology takes a large leap from IS-95A and IS-95B. The number of Walsh codes available has been increased to 128; there has also been an increase in the length of the Walsh codes. This is coupled with several performance improvements, including stronger error coding, faster and more improved forward power control, and transmit diversity. The end result is the ability to possibly double voice capacity, as well as offer high-speed packet-switched data services (up to 153 Kbps). There are several other improvements as well, including several new physical channels that will improve performance (e.g., the quick paging channel, which can increase the handset's idle time substantially—which would mean longer battery lives).

The 3X technology would take another leap from 1X. The concept is basically to use three 1.2288-MHz channels together (hence, it uses what is called *spreading rate 3*) (see Figure 8.2). In 3X, the Walsh codes can be up to 256 chips long, allowing for much more voice capacity in addition to very-high-speed data rates (perhaps up to 2 Mbps). In addition, 3X will have some degree of world standardization, as far as the core network it works off. It will be able to operate on ANSI-41, which is the

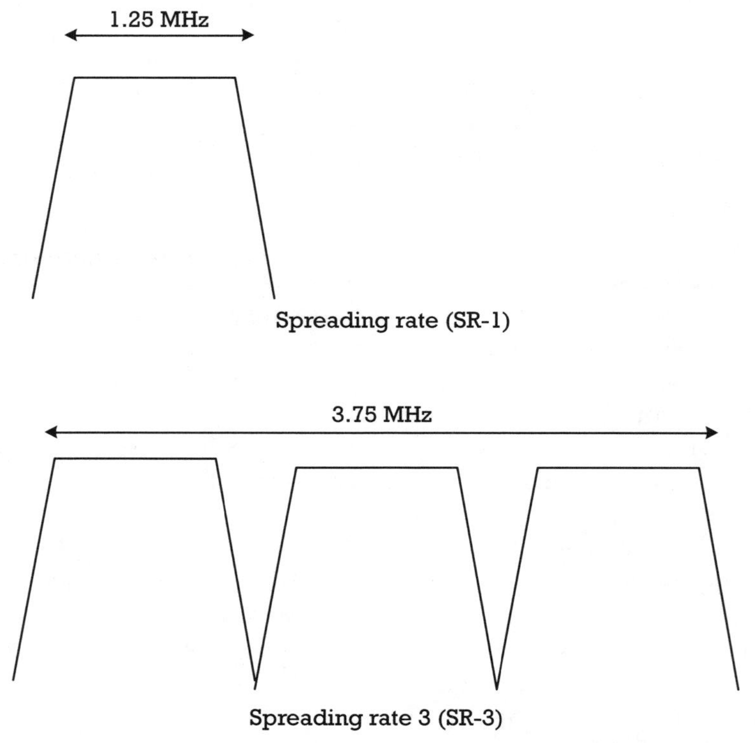

Figure 8.2 The 3X technology uses three 1X carriers; hence, it uses spreading rate 3.

primary core network many American TDMA and CDMA networks operate on today, or GSM-MAP, the core network more standard throughout the world for GSM and its future upgrade paths. The 3X business case, however, may result in this technology not being widely deployed. While it may offer higher data rates, the cost of deployment may be quite high, and 1X performance may be enough for some time. For that reason, most of this text will focus on 1X technology—although the 3X is not terribly different.

8.4 Radio Configurations

In CDMA2000, operators have quite a bit of flexibility in the services they can offer via various radio configurations. Both the forward and reverse links have various radio configurations that can be deployed, which will provide varying levels of data rates, and therefore services, that they can support.

As discussed in Section 8.2, the two 3G versions of CDMA are often referred to as 1X and 3X. The 1X version is said to use a *spreading rate* (SR) of 1 (SR1), while 3X using a *spreading rate of 3* (SR3).

You will recall from the previous chapter on CDMA that the IS-95 spreading rate is 1.2288 Mcps. This spreading rate, 1.2288 Mcps, remains the same in 1X. In 3X, naturally, it is tripled, so the spreading rate becomes 3.6864. If the spreading rate remains the same, but the effective data rate increases—as is the case with the higher speed data transmissions—the processing gain decreases. This means that performance improvement measures need to be used to boost the quality to ensure adequate reception and transmission, which is one of the driving forces of the 3G standard.

In the end, there will be nine different radio configurations for the forward link in CDMA2000. These are divided into three groups: radio configurations 1 and 2 are IS-95B configurations, 3 through 5 are 3G-1X configurations, and 6 through 9 are 3G-3X configurations. As you can see in Table 8.1, the different radio configurations enable different data rates, modulation, and encoding. Naturally, the handsets must also be compatible with these data rates in order to make use of the capabilities.

In the reverse link, there are six different radio configurations. The first two are IS-95 compatible, with the same spreading modulation. RC2 uses enhanced convolutional encoding, which allows for higher data rates. The last four, however, are dramatically different. The reverse link in CDMA2000 has many changes aimed at increasing capacity as well as enabling simultaneous voice and data channels. RC5 and RC6 are designed for the spreading rate 3, while RC1 through RC4 are designed for spreading rate 1.

A very nice aspect of CDMA2000 is that the reverse link and forward link radio configurations can be independently mixed and matched. This can allow for a full-

Table 8.1 Downlink Radio Configurations in CDMA2000

Radio Configuration	*Spreading Rate*	*Data Rates (bps)*
1 (IS-95B)	1	1,200; 2,400; 4,800; 9,600
2 (IS-95B)	1	1,800; 3,600; 7,200; 14,400
3 (3G-1X)	1	1,200; 1,350; 1,500; 2,400; 2,700; 4,800; 9,600; 19,200; 38,400; 76,800; 153,600
4 (3G-1X)	1	1,200; 1,350; 1,500; 2,400; 2,700; 4,800; 9,600; 19,200; 38,400; 76,800; 153,600; 307,200
5 (3G-1X)	1	1,800; 3,600; 7,200; 14,400; 28,800; 57,600; 115,200; 230,400
6 (3G-3X)	3	1,200; 1,350; 1,500; 2,400; 2,700; 4,800; 9,600; 19,200; 38,400; 76,800; 153,600; 307,200
7 (3G-3X)	3	1,200; 1,350; 1,500; 2,400; 2,700; 4,800; 9,600; 19,200; 38,400 76800, 153600, 307200, 614400
8 (3G-3X)	3	1,800; 3,600; 7,200; 14,400; 28,800; 57,600; 115,200; 230,400; 460,800
9 (3G-3X)	3	1,800; 3,600; 7,200; 14,400; 28,800; 57,600; 115,200; 230,400; 460,800; 518,400; 1,036,800

blown 3G-3X forward link coupled with a 3G-1X reverse link. The idea is that Internet browsing requires much faster downlink speeds versus uplink speeds, and this would be a much more economically attractive method of deployment. Specifically, however, if a mobile is able to receive the forward link from the base station in RC6 or RC7, it must be able to transmit using the reverse configurations of RC3 or RC5 (see Tables 8.1 and 8.2).

8.5 Spreading and Channel Coding Changes

CDMA2000 introduces several new methods of coding and spreading that enable stronger physical links and, thus, higher capacities and better performance.

The most obvious changes deal with the Walsh codes. In IS-95, Walsh codes are fixed at 64 bits; thus, there are 64 codes available per carrier. In 1X, Walsh codes are doubled, allowing for 128 codes. In 3X, 256 Walsh codes are available. This is enabled by changes in the spreading modulation method used. The details of the spreading modulation are a bit complex and above the intended scope of this text, but suffice it to say that IS-95 uses a two-level spreading channel versus the CDMA2000 channel, which uses four channels.

The concept of variable-length Walsh codes deals with data transmission. By being able to use varying length Walsh codes, the bandwidth for data services can be varied accordingly.

What is important to understand, from a practical perspective, is that the 1X and 3X Walsh codes are derived from the original 64 (if you go back to the earlier

Table 8.2 Uplink Radio Configurations in CDMA2000

Radio Configuration	*Spreading Rate*	*Data Rates (bps)*
1 (IS-95B)	1	1,200; 2,400; 4,800; 9,600
2 (IS-95B)	1	1,800; 3,600; 7,200; 14,400
3 (3G-1X)	1	1,200; 1,350; 1,500; 2,400; 2,700; 4,800; 9,600; 19,200; 38,400; 76,800; 153,600; 307,200
4 (3G-1X)	1	1,800; 3,600; 7,200; 14,400; 28,800; 57,600; 115,200; 230,400; 460,800
5 (3G-3X)	3	1200; 1,350; 1,500; 2,400; 2,700; 4,800; 9,600; 19,200; 38,400; 76,800; 153,600; 307,200; 614,400
6 (3G-3X)	3	1,800; 3,600; 7,200; 14,400; 28,800; 57600; 115,200; 230,400; 460,800; 1,036,800

chapters, you can see exactly how the Walsh codes are derived and how a Walsh code can really be any length). In other words, one IS-95 Walsh code (annotated as W^{64}) can create two 1X codes (annotated as W^{128}), and four 3X codes (annotated as W^{256}). This means that if you use W^{64}_{6} (the 6th code of W^{64}) in a 1X system because the user's phone is not 1X compatible, you actually will be using two 1X Walsh codes and four 3X Walsh codes (which would then be unavailable for use). One can easily see, then, that while a 1X network would be backward compatible, the benefits (in terms of capacity) would quickly diminish if an operator does not deploy 1X mobiles quickly enough in volume. It also means that in the same CDMA carrier, you can have IS-95 Walsh channels (using 64-bit Walsh codes) and 1X Walsh channels (128 bits) simultaneously. Obviously, older IS-95 phones would only be able to use the W^{64} encoded channels, while the CDMA2000 phone could use them all. Figure 8.3 shows a WCDMA spreading tree.

In IS-95, the set limit of 64 Walsh codes was an absolute limit (i.e., one CDMA carrier could not have any more than 64 Walsh codes at any time). CDMA2000 has the limits of 128 or 256; however, these are not hard limits. Quasi-orthogonal Walsh code multiplication enables an RF channel to have more than the set limit of Walsh codes by applying a complex mask to the Walsh codes, creating *quasi-orthogonal functions* (QOFs). The QOF can be used for "extra" channels as needed.

CDMA2000 also employs some stronger error encoding methods as well. This strengthens the links by improving the coding gain as the links are received. Additionally, for high-speed data, *turbo codes* are utilized. Essentially, this is a method of convolutional encoding that allows for even higher coding gains that can be needed for data transmissions.

Just as in IS-95, interleaving and symbol repetition, discussed in earlier chapters, is used to facilitate the various data rates as well as improve performance of the transceivers.

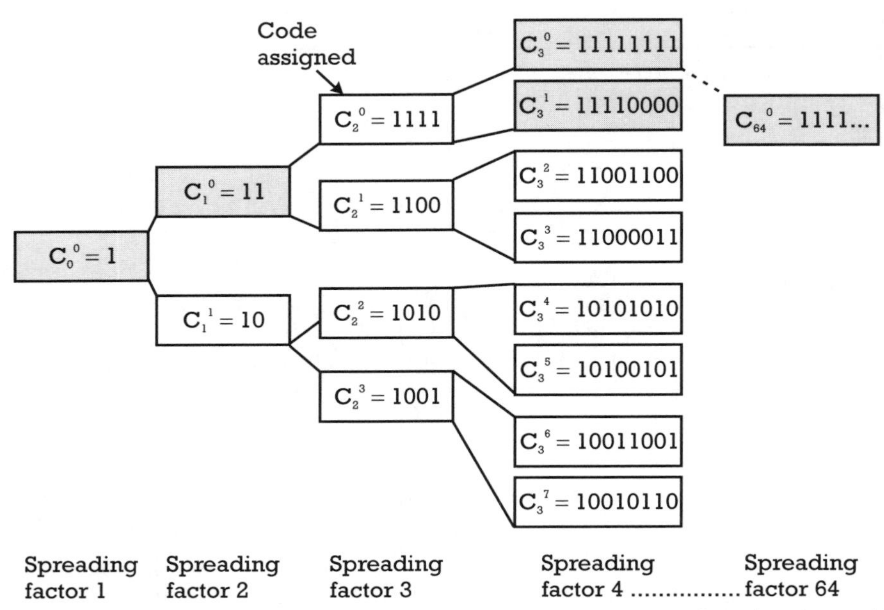

Figure 8.3 A spreading code tree. The codes in the darkened boxes are all rendered unusable once the assigned code is used. This is due to the unusable codes being either roots or being derived from the assigned code.

8.6 Air Interface Differences

You should remember from previous chapters that IS-95 used QPSK modulation on the forward link. CDMA2000 also uses QPSK, but you might hear high-level engineers call it "true" QPSK. The details are a bit above the scope of this text; however, the QPSK used in IS-95 is effectively a dual-BPSK with orthogonal diversity, because the data streams in the I and the Q channels are identical (see Figure 8.4). In CDMA2000, the I and the Q channels are different. The result is that the same amount of chips require half as much bandwidth. The downside, of course, is that it is more sensitive to phase errors. However, it has several other improvements in coding, as discussed earlier, as well as the new forward power control and forward transmit diversity, which will be discussed later. This can be mitigated, thus allowing a doubling of spectrum efficiency.

On the reverse link, there are quite massive changes. As you will see in the next section, the reverse link in CDMA2000 adds several different physical channels. Similar to the forward link, it adds a pilot channel, as well as supplemental channels (for data) and control channels for signaling information. This is obviously very different than IS-95, where only one physical channel was transmitted. Similar to the forward link, the reverse link uses Walsh functions to differentiate the different physical channels it can transmit.

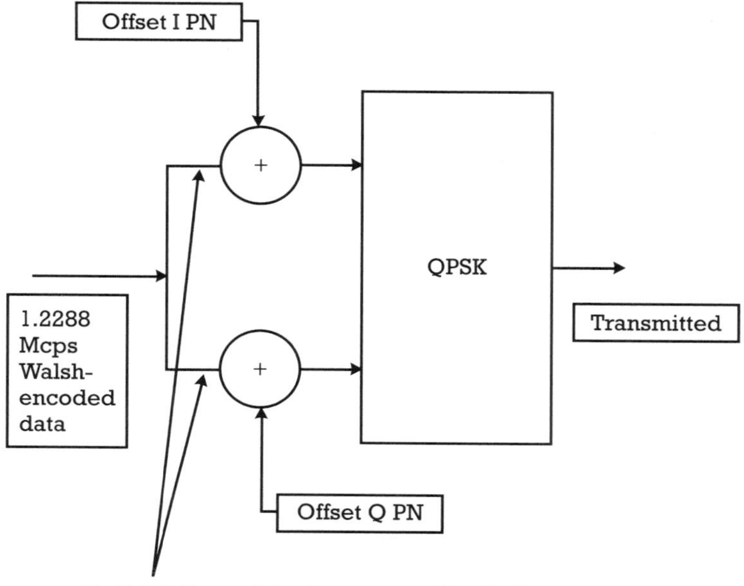

Figure 8.4 CDMA2000 uses "true" QPSK, in that the I and Q channels contain independent data.

You will remember from previous chapters that the reason O-QPSK is used on the reverse link is to prevent zero crossings, which reduces the peak-to-average power ratio, making handset designs easier. In CDMA2000, however, the reverse link has multiple channels. O-QPSK would not, in this case, prevent zero crossings and thus is not good for reverse link modulation. Instead, a modulation format called *hybrid phase shift keying* (HPSK), also known as *orthogonal complex quadrature phase shift keying* (OCQPSK), is used.

HPSK is a very complex modulation scheme and involves several very complex steps. To start with, the multiple channels that are to be transmitted are assigned either the I or the Q modulator path. Because these different channels might have varying power levels, an extremely complex scrambling process is used, in addition to the Walsh function spreading. This is actually called *orthogonal variable spreading function* (OVSF). It identifies each channel. In the end, all of the channels can be modulated on the same carrier, with a low probability of zero crossings.

As you can follow in Figure 8.5, the channels are spread with the appropriate Walsh *cover* sequence. They are then given gain as needed and summed together. This then creates the I and Q data stream. This stream is then spread by the *complex multiplier,* which includes spreading by the long code with the mobile-specific long code mask (same as in IS-95). Once the sequences come out of the complex

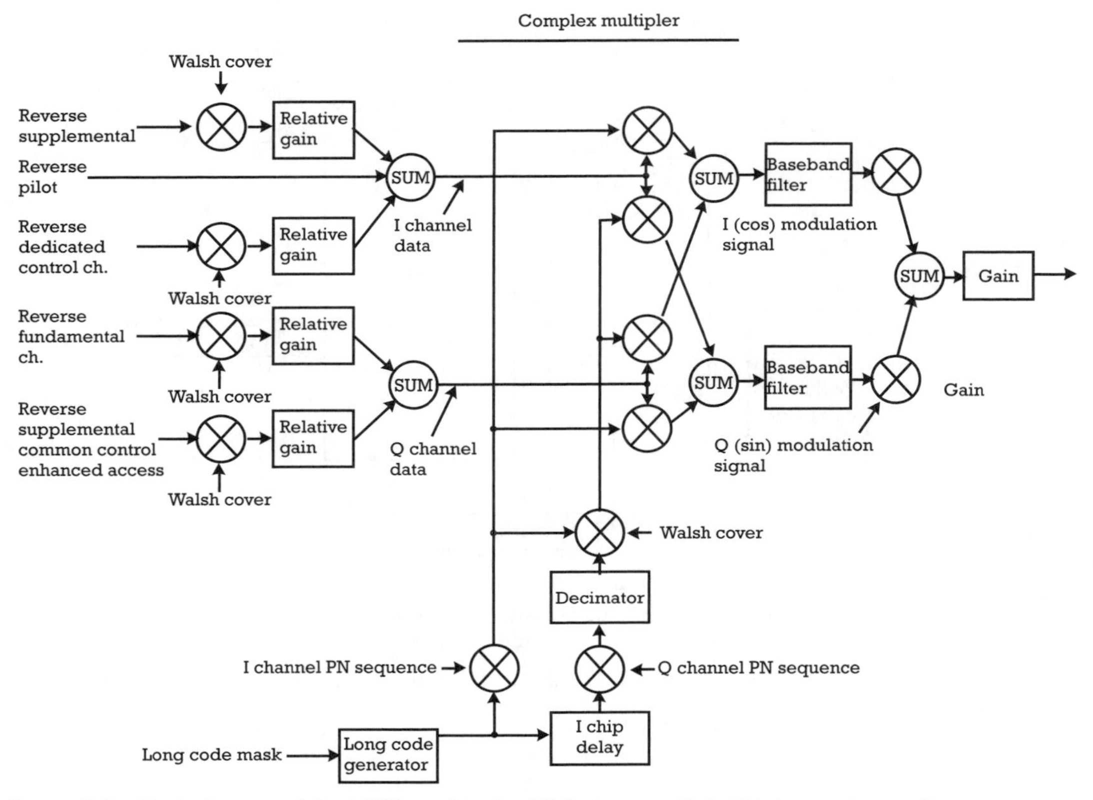

Figure 8.5 Block diagram of the HPSK used in the RC 3–4 reverse link. This is a very complex process.

multiplier, the sequences can then be I/Q modulated. The end result is that the constellation diagram will look similar to QPSK but with very limited zero crossings.

Obviously, this text has attempted to simplify what is a very complex process. At the technician level, however, the specifics of the process of HPSK are not terribly important. What is important is to understand the concepts of multiple channels in the reverse link, that the link uses a masked long code similar to IS-95 for security, and that *error vector magnitude* (EVM) can still be used as a measurement of the quality of modulation.

8.7 Power Control Changes

One of the more important changes in CDMA2000 is the introduction of fast power control on the forward link. In IS-95, the individual code channel power on the forward link was varied based on received messages from the mobile indicating the quality of the link. This enables the base station to vary its power upon receipt of the message, every 20 ms at the most. This is obviously much slower than on the reverse link, where power can be varied every 1.25 ms.

The reasoning for this was that the reverse link transmits *many to one* base station, and obviously the near-far issue would cause a lot of problems if power was not adjusted quickly. The forward link is *one to many*, and thus the near-far issue is not as big a concern. That said, there is still mutual interference in the forward link (all the channels do still transmit in the same frequency, albeit with true orthogonality) as well as fast fading situations that require better power control to maintain quality. CDMA2000 rectifies this issue by implementing the new power control process.

The forward link fast power control is implemented in a manner very similar to the reverse link power control. It consists of an outer loop, which sets the desired E_b/N_0 in order to attain a certain FER. The inner loop then measures the received power in order, sending power control bits back to the base station via punctured bits in the power control subchannel on the newly implemented reverse pilot channel. This enables a fast power control with speeds up to the same as in the reverse link—every 1.25 ms, or 800 times a second.

Power control in the reverse link is obviously still extremely important. It has a direct impact on the capacity of the network; the more accurate power control is, the better the performance of the system will be. Thus, in CDMA2000, the power can be adjusted in finer amounts than before—as low as 0.25-dB increments.

There is also a change in where the reverse power control bit can come from. While on a traffic call, it can get the power control bit from the forward traffic channel (now called the *forward fundamental channel*), as it does in IS-95, or it can find the bit in the new dedicated control channel (discussed in the next sections).

You will also see in Chapter 9 that the reverse link in CDMA2000 now has multiple channels from each mobile. These other channels require power control. Two of these channels, the enhanced access channel and the reverse common control channel, receive their power control instructions from a new forward link channel: the forward common power control. In addition, as with the forward link, there is now the concern of code domain power control (i.e., ensuring the various reverse code channels are transmitted at appropriate levels relative to each other). Like the forward link, these code power levels are determined relative to the pilot channel.

8.8 Physical Channel Overview

As has been discussed, CDMA2000 is a backwards-compatible system. That means that an IS-95 phone and a 1X phone would be able to access a CDMA carrier at the same time and complete a conversation. With all of the differences described in the previous sections, this might seem rather difficult, but it comes together nicely in practice.

The standards body that developed CDMA2000 developed standard naming conventions for the channels, which this text will follow. Upper case letters are for physical channels, while lower case letters are used for logical channels. The first letter of both logical and physical channels indicates whether it is a forward or reverse channel, or both. Table 8.3 shows all of the CDMA2000 physical channels and their abbreviations.

A CDMA2000 forward link will have all of the same channels as an IS-95 channel. The pilot is found on W_0^{64}, the sync remains at W_{32}^{64}, and the paging channels can be W_1^{64}–W_7^{64}, just as in IS-95. New physicals channels, however, will make use of the new 128-bit-length Walsh codes. Thus, to an IS-95 mobile, the signal will be completely normal (as far as IS-95 signals go), but if the mobile has 1X capabilities, it will be able to make use of the new channels and changes that increase the performance of the network. Again, it is critical to recognize that the key to realizing the benefits of CDMA2000 is to deploy the CDMA2000 handsets.

Other forward channels that have assigned Walsh codes include the quick paging channels, which use W_{80}^{128}, W_{48}^{128}, and W_{112}^{128}; the 3X pilot, which uses W_0^{256}; and the orthogonal transmit diversity pilot, which uses W_{16}^{128}. Auxiliary pilots, which are used in smart antenna applications, make use of the QOFs discussed earlier.

In the reverse link, there are several new channels, all of which are identified via Walsh functions of varying lengths (due to the varying data rates of the channels). These new physical channels and their abbreviations are shown in Table 8.4.

Table 8.3 CDMA2000 Physical Channels and Their Abbreviations

Channel Name	*Physical Channel*
F-FCH	Fundamental channel
F-DCCH	Dedicated control channel
F-SCCH	Supplemental code channel
F-SCH	Supplemental channel
F-PCH	Paging channel
F-QPCH	Quick paging channel
F-CCCH	Common control channel
F-PICH	Pilot channel
F-APICH	Dedicated auxiliary pilot channel
F-TDPICH	Transmit diversity pilot channel
F-ATDPICH	Auxiliary transmit diversity pilot channel
F-SYNCH	Sync channel
F-CPCCH	Common power control channel
F-CACH	Common assignment channel
F-BCCH	Broadcast control channel

Table 8.4 New Physical Channels and Their Abbreviations in the Reverse Link

Channel Name	*Physical Channel*
R-FCH	Fundamental channel
R-DCCH	Dedicated control channel
R-SCCH	Supplemental code channel
R-SCH	Supplemental channel
R-CCCH	Common control channel
R-PICH	Pilot channel
R-ACH	Access channel
R-EACH	Enhanced access channel

8.9 Forward Physical Channels

Here is an overview of the forward physical channels (see Figure 8.6):

- *Forward pilot channel (F-PICH):* This channel remains the same in CDMA2000 as it was in IS-95. It is an unmodulated channel (no data) that uses Walsh code 0 (all zeros). Because it is a backward-compatible channel, it uses the 64-bit Walsh code.
- *Forward transmit diversity pilot channel (F-TDPICH):* Transmit diversity will be discussed in a different section. Essentially, it involves splitting the bits in the forward channels and transmitting the forward channel on two different antennas. The F-TDPICH provides the timing reference for the second-antenna diversity signals. Like the F-PICH, it is an unmodulated channel that contains no data. It is spread with W_{16}^{128}, so it is not a backward-compatible channel, and the phone must be 1X capable to use it. The F-PICH remains the primary pilot as far as setting the cell radius, so cells should be designed such that the F-RDPICH does not overpower the F-PICH. The F-TDPICH can be set at 0, –3, –6, or –9 dB down from the F-PICH.
- *Forward dedicated auxiliary pilot channel (F-APICH):* This is used for smart-antenna applications. In smart-antenna systems, a sector can be further divided in very small *beams* in order to eliminate interference and improve the forward and reverse link. Each beam serves essentially as its own sector and thus requires its own pilot. In such configurations, it is possible to use many more code channels than even the 128 available in 1X; this is one of the reasons for having QOFs available. The F-APICH can use a Walsh function or a QOF. As there obviously could be several F-APICHs per carrier per sector—which would also demand more code channels for more traffic channels—QOFs will most likely be used for these rather than Walsh functions.

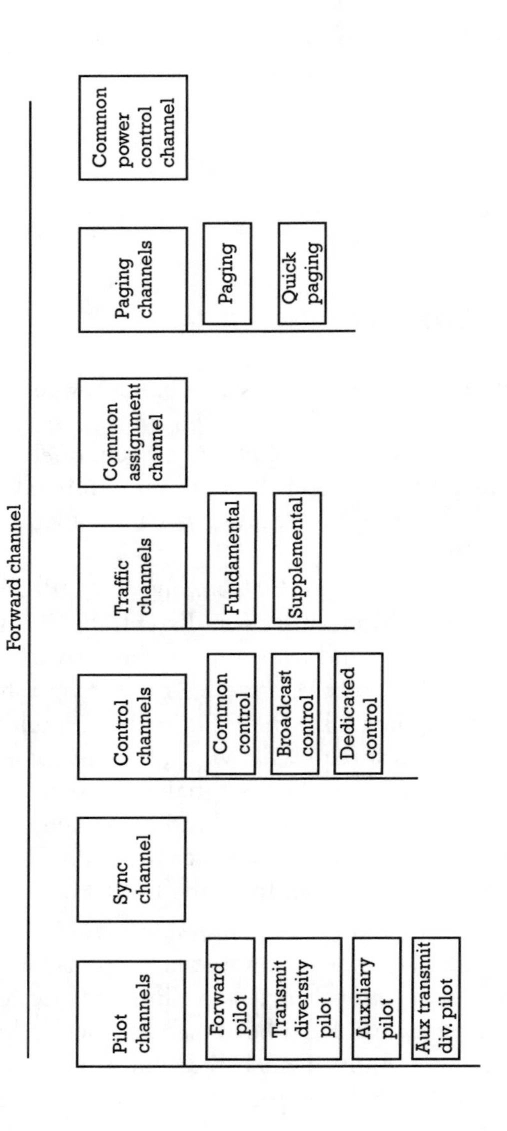

Figure 8.6 Structure of the forward channel in 1X.

- *Forward auxiliary transmit diversity pilot channel (F-ATDPICH):* If you understand that each beam in a smart-antenna system acts as its own sector, and you also wish to use transmit diversity in each beam, then you will need to transmit the F-ATDPICH. It serves the same function as the F-TDPICH, except it will be assigned a different Walsh or QOF function.
- *Forward sync channel (F-SYNCH):* This remains the same as in IS-95. A bit more information needs to be communicated in the F-SYNC in CDMA2000, as the mobile needs to know if the base station is 1X or 3X capable. As in IS-95, the F-SYNC's frame and interleaver block and frame are time aligned exactly with the F-PICH. This means that once the mobile time aligns with the pilot, it can easily just switch to W_{32}^{64} and read the sync message, which is the same for all mobiles.
- *Forward paging channel (F-PCH):* F-PCH also keeps the same physical structure as in IS-95. The difference, however, is the addition of several other channels that can assist in transmitting overhead and direction messages (or logical channels) to specific mobiles, which can increase the performance of the paging channels. In addition, the sleep mode has been improved with the use of the quick paging channel (discussed next). In general, the paging and messaging function while not on a call is accomplished with four channels, with a fifth optional: the *forward quick paging channel* (F-QPCH), the *forward common control channel* (F-CCCH) (see Figure 8.7), the *forward broadcast control channel* (F-BCCH), and the F-PCH. Optionally, the *forward common assignment channel* (F-CACH) can also be used.
- *F-QPCH:* You should remember from the previous chapters the basic concepts behind the sleep mode. The purpose of the sleep mode is to reduce the amount of time the mobile has to stay active, thus increasing the battery life. In IS-95, the mobile simply woke up at set points to check a slot on the paging channel to see if it should wake up. In CDMA2000, it is similar, except that the F-QPCH is dedicated to this function. Each mobile has a specific slot assigned to it to monitor. It is time aligned at a specific point in the pilot sequence for easy acquisition. There are three indicators a mobile can wake up to monitor. This first is the paging indicator, which tells the mobile to read the next slot in either the paging channel or the F-CCCH, if used. The configuration change indicator lets the mobile know it needs to get new configuration settings (settings are found in the CONFIG_MSG_SEQ message). Finally, the broadcast indicator tells the mobile it needs to monitor the broadcast message slot on the F-CCCH. Implementation of the F-QPCH has an enormous impact on CDMA2000, possibly reducing battery consumption by a factor of as much as 15 during sleep mode. In IS-95, the ratio of up time to sleep time is 13:1, versus the comparable ration in 1X, which will be 200:1 (see Figure 8.7).

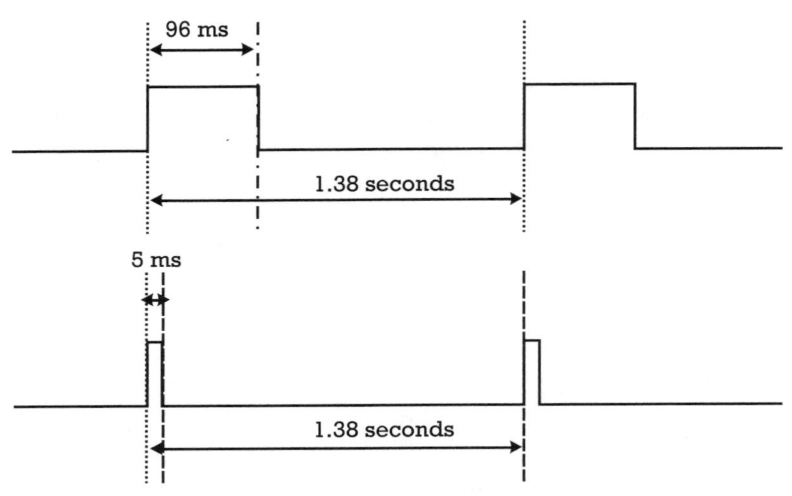

Figure 8.7 The use of a separate quick paging channel allows the mobile to "wake" for a significantly shorter period of time, as the slot it needs to monitor is only 5 ms, versus the 96 ms it needed to be awake for in IS-95.

The F-QPCH can be transmitted on three 1X channels: W_{80}^{128}, W_{48}^{128}, and W_{112}^{128}.

- *F-CCCH:* The F-CCCH works with the F-PCH to transmit mobile-specific messages. Like the F-PCH, the F-CCCH is divided into slots that can be assigned to specific mobiles to monitor. The data rate on the channel can vary between 9,600, 19,200, and 38,400 bps. When not on a voice or data call, mobile-specific messaging and pages are sent on the F-CCCH.
- *F-BCCH:* The F-BCCH is used to transmit messages that go to all of the mobiles in the area covered by the cell. Messages might include broadcast short messages (advertisements or news) as well as static paging messages. The channel can be transmitted at 4,800, 9,600, or 19,200 bps.
- *F-CACH:* The F-CACH is an optional channel that is used to assign reverse channels for the mobile to use in the enhanced access channel procedure (discussed separately). It is transmitted at 9,600 bps.
- *Forward common power control channel (F-CPCCH):* When not on a voice or data call, the mobile still needs to receive the bits for its closed-loop power control. Each mobile is assigned to watch one bit in this channel, which tells it whether to increase or decrease its power. The bits are not encoded but simply multiplexed, which allows for fast reception by the mobile required to make the fast power control work. Multiple F-CPCCHs can be used per carrier as needed.
- *Forward fundamental channel (F-FCH):* The F-FCH is the primary carrier of voice or data information in the traffic channel. The channel is the same as in

IS-95, depending on the radio configuration. There is one F-FCH per traffic channel (a traffic channel can consist of multiple code channels to provide the data-rate service required). Various radio configurations do have variable data rates, but constant spreading rates. This is achieved by the different levels of channel coding or symbol repetition.

- *Forward supplemental code channel (F-SCCH):* Up to seven F-SCCHs can be used in a traffic channel. The F-SCCH is only used in radio configurations 1 and 2, and it can carry either data or voice data.
- *Forward supplemental channel (F-SCH):* The F-SCH is used in radio configurations 3–6 for higher data-rate service. The F-SCH can use the variable-length Walsh codes to ensure the spreading rate remains constant. Up to two F-SCHs can be used in a traffic channel.
- *Forward dedicated control channel (F-DCCH):* You will remember from IS-95 that any signaling that needed to be sent to the mobile during a voice call was sent on the forward traffic channel, mixed in with the voice data. CDMA2000 improves this situation by using a dedicated control channel just for this data. One F-DCCH is used with each traffic channel. This then frees up the F-FCH to transmit only actual user data as opposed to the forward signaling information, thus significantly improving performance.

8.10 Reverse Physical Channels

Here is an overview of the reverse physical channels (see Figure 8.8):

- *Reverse pilot channel (R-PICH):* New in CDMA2000, the R-PICH provides a timing reference that makes the reverse link coherent and thus perform substantially better. Power control for the forward link also uses the R-PICH. While adding the pilot in the reverse link obviously will increase the battery consumption, as well as add additional interference, the benefits far outweigh the drawbacks, as initial acquisition, time tracking, reception by the forward link, and power control all receive large performance improvements.
- *Reverse access channel (R-ACH):* The access channel is a backward-compatible channel, hence it works as it did in IS-95. You will remember from IS-95 that the R-ACH transmits bursts, or probes, during specific time slots.
- *Reverse enhanced access channel (R-EACH):* The R-EACH performs the same function as the R-ACH but has several enhancements (hence the name). There are two modes of enhanced access supported. The first, basic access mode, is essentially the same as in IS-95. In this mode, the R-EACH carries the access messages (origination requests and page responses). The second mode is the reservation access mode. This is used to reserve certain radio resources,

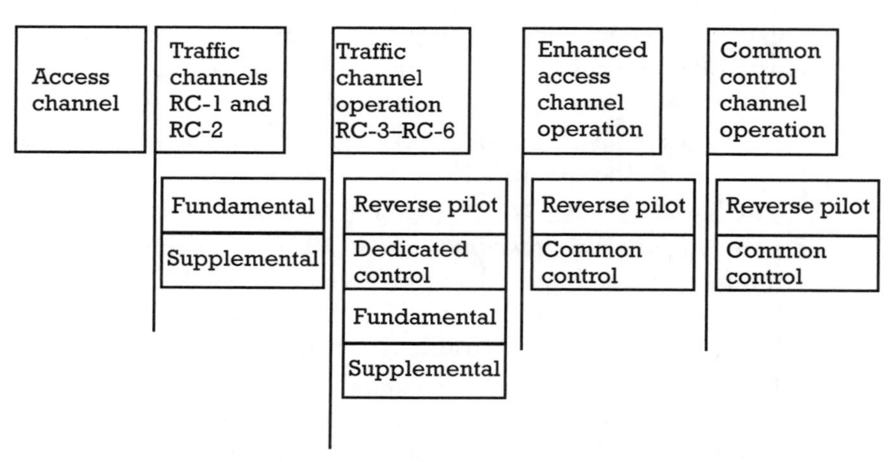

Figure 8.8 The reverse channel structure in 1X.

such as Internet access. In this mode, the R-EACH is used primarily to establish the access, and the actual message is carried by the *reverse common control channel* (R-CCCH). Both of these modes make use of the R-PICH to initiate the access attempt. Prior to the access attempt, the reverse pilot power is increased 6 dB as an R-EACH preamble. This helps the base station receive the message and sync up perfectly, thus improving access performance.

- *R-CCCH:* The R-CCCH is used in conjunction with the R-EACH in the reservation access mode. It is used for transmission of user and signaling information to the base station when not in a voice or data call. It comes into play quite a lot for mobile applications with "always on" Internet access, where the channel is used to request packet access. The R-CCCH is closed loop power controlled, as described earlier.
- *Reverse dedicated control channel (R-DCCH):* Similar to the F-DCCH, the R-DCCH is used to transmit signaling and user information during a call. The channel transmits only if there is data to be sent; thus, the decision to transmit is made on a frame-by-frame basis.
- *Reverse fundamental channel (R-FCH):* The R-FCH is used for transmission of voice or data back to the base station. The R-FCH can use 5-ms or 20-ms frame periods, which allows for lower data transmissions without repeating transmissions.
- *Reverse supplemental code channel (R-SCCH):* The R-SCCH is similar to the F-SCCH, in that it is used in radio configurations 1 and 2 to increase the data services available. Up to seven can be used per traffic channel.

- *Reverse supplemental channel (R-SCH):* Again, similar to the F-SCH, the R-SCH is used to offer high-speed data service in radio configurations 2 through 6. Up to two F-SCHs can be used in a traffic channel. Like the forward link, the R-SCH uses variable Walsh codes to ensure the spreading rate remains the same with the variable data rates.

8.11 Logical Channels and Layering Overview

Section 8.10 described the physical channels in CDMA2000. These physical channels serve as the final layer of processing to get the user data and control information to and from the base station and mobile. This data is transported via logical channels that can be placed on one or more physical channels. One aim for the future is to be able to standardize this logical layer, such that there is a degree of standardization and open interface from the base station backward towards the switch. By separating logical channels from the physical channels, it becomes possible to harmonize different standards, something the world bodies have viewed as a goal.

Thus, the logical channels are not sensitive to the radio characteristics of the physical channel; they are essentially transport channels for the data that can be plugged into the various physical channels as needed. These logical channels and messages are often called the layer 2 and layer 3 messages. Layer 2, called the *link layer,* includes the *link access control* (LAC) sublayer, which packages data for point-to-point communication, and the MAC sublayer, which controls the access to the physical layer. In essence, LAC gets the upper level messages ready to be transmitted, while MAC uses suitable methods to actually deliver the messages (via the physical layer).

Layer 3, the layer above the LAC and MAC, are the actual data and control messages themselves. Many of these messages are described in the IS-95 chapter.

There are several new messages and settable parameters in CDMA2000 to support the new features and procedures. Several services then would need to be either established or negotiated, usually at the start of a call. These services might include radio configurations and traffic channel transmission rates. In addition, several parameters would need to be transmitted to the mobile as it establishes a call on a base station, such as power control parameters, various variable-rate information, and the configuration of the R-PICH.

Important new messages in CDMA2000 include a new forward link power control message, which sets the "outer loop" of the forward power control. Another important set of messages deals with establishing and maintaining data service on the SCH. There are messages to request SCH operation as well as assign the channels.

8.12 Handoff Changes

While, for the most part, the basic concept of handoffs in CDMA2000 work the same as in IS-95, some changes are designed to limit the amount of handoffs that take place. This is desirable because each soft handoff–active PN uses a radio element. In many instances in IS-95A networks, this overuse of soft handoffs is what limited the capacity of the network, because there would be no forward channel elements available for us. The implementation of the dynamic soft handoff thresholds took place in IS-95B, which was not deployed in the United States but was deployed in Asia.

To lower the amount of handoffs taking place, two dynamic and two static thresholds are used. The first threshold, T1, is a static threshold used to determine whether a pilot should be moved to the candidate set. Once in the candidate set, a second dynamic threshold is used, T2. T2 is a function of the total pilot energy that is demodulated coherently. If the candidate is higher than T2, it is added to the active set in a soft handoff. The idea is that handoffs only occur when they have to—or when adding the neighbor pilot would be beneficial.

A third threshold is also dynamic, T3. T3 is similar to t_drop, but it is dynamic in that it is based on the total pilot energy in the pilot set. If the pilot drops below T3 for a set amount of time, the pilot will drop back to the candidate set. T4, the final threshold, is static and is used to determine the level at which a pilot is dropped altogether.

These handoff thresholds will have a large impact on capacity, and they may increase the demand and strain placed on the reverse link. You will remember that in soft handoff, the reverse link is transmitted by one mobile and is actually helped due to the diversity of being received by two sectors. The forward link, however, uses a code channel from two base stations to transmit. Being limited by the amount of radios installed, and ultimately the amount of available Walsh codes, networks with high handoff factors generally are forward-link limited. Thus, as handoffs come down, the reverse link begins to take the strain, and often the reverse link begins to limit the capacity.

8.13 Forward Transmit Diversity

Trying to give end users high data rates at varying distances from cell sites is a very complex problem. You will recall that with the high data rates, the processing gains go down. This means the link gets weaker and requires either more power or better sensitivity to maintain quality. Another alternative, however, is the use of transmit diversity. Today, almost all base stations make use of receive diversity (i.e., they use two receive antennas and either combine the signals or choose the path with the best reception). The idea with forward transmit diversity is somewhat similar. The base station will use two transmit antennas to transmit the forward link. There are two methods used, both of which are optional, and either can be used, depending on the base station manufacturer.

Space-time spreading (STS) transmits all of the forward channel data on both antennas but spreads the symbols with different orthogonal, or quasi-orthogonal, functions. The mobile can then receive both paths and combine them for a more robust reception.

Orthogonal transmit diversity (OTD) distributes the forward channel symbols among the diversity antennas and then spreads the symbols with different orthogonal or quasi-orthogonal functions. In this way, the mobile receives both signals and combines them for complete reception.

8.14 TDD Mode

The 3G standards bodies were tasked with standardizing TDD versions of their formats. Thus, it is possible that some carriers might deploy the TDD version of CDMA2000. Widespread deployment is not expected. Essentially, the system involves sharing time between the receive and transmit, in order to make use of less bandwidth. Because of this time sharing, however, the system would obviously have some performance drawbacks.

8.15 Conclusion

CDMA2000 is a fairly extensive overhaul of IS-95. It includes several new channels on the forward link, very different reverse link channels, and different spreading and modulation techniques and procedures. The end result is a system that will substantially increase voice capacity while offering high-speed data services that many predict will catch on soon. Considering that the format is an upgrade and can often be implemented by making use of existing hardware, it is assured that some forms of CDMA2000, most likely 1X, will be deployed in almost all existing IS-95 networks.

CHAPTER 9

WCDMA

9.1 Introduction

WCDMA technology has many similar features to CDMA2000. While much of the nomenclature is different, the overall concepts are quite alike. It is for this reason that it is strongly recommended that the reader go through the earlier chapters on CDMA and CDMA2000 prior to trying to learn the WCDMA system, as many of the basic concepts will not be reviewed in detail in this chapter, especially the topics of code othogonality, spreading, convolutional encoding, PN codes, Rake receiving, and transmit diversity.

There are some differences between CDMA2000 and WCDMA that will have a large impact as the systems are rolled out in the field. The largest difference is that CDMA2000 air interface has the added task of being fully backward compatible, while WCDMA was developed without this in mind, for the sole purpose of being a "true" third generation mobile system right from the start. This has both an advantage and a disadvantage. The advantage stems from the system not being encumbered with the added baggage associated with 2G technology; it does not need to deal with legacy technology. The disadvantage lies in the cost and complexity of rolling such a system out, as it will require a Greenfield deployment. In addition, as WCDMA will more than likely be rolled out over systems currently using TDMA technologies (IS-136 or GSM), engineering teams will have to learn an entirely new technology.

9.2 History of WCDMA

In the 1990s, wireless executives began to see that 2G systems would need to be upgraded or replaced to provide more features, services, and efficient use of spectrum. As early as 1992, the *World Administrative Radio Conference* (WARC) began work to define new spectrum that would be used for this purpose. This might be surprising, considering in 1992 GSM, IS-95, and IS-136 were still in early deployments. Around the same time, the *International Telecommunication Union* (ITU), a collection of scientists and executives focused on standardization, began work on defining what the 3G system should be able to do (see Figure 9.1). Out of this group came the *International Mobile Telecommunications System 2000* (IMT-2000) system, also known as the UMTS. Essentially, UMTS defined a wireless mobility system that would allow high-speed communication services up to 2 Mbps.

Thus began a process of developing the new technology and standards that would make up the new system, with the hope of a global standard to allow for a high degree of economy of scale to keep costs down and allow for easy international roaming. A global committee was formed to coordinate this process, which was known as the *Third Generation Partnership Program* (3GPP). As time went on, it was clear there would be more than one new air interface. Thus a new partnership program was set up to ensure the CDMA2000 standard would meet the IMT-2000 standard, alongside the original partnership that was working on what was to become the WCDMA system. Thus, you may have heard reference to 3GPP1 and 3GPP2, with 3GPP1 referring to WCDMA, and 3GPP2 referring to CDMA2000. In addition, it would now appear a third 3G format, known as *time division–synchronous code division multiple access* (TD-SCDMA), will also gain market share, as it is being deployed in China as a 3G system.

The first deployments of WCDMA have taken place in Japan, which activated the first commercial system in 2001. Already, over 75 countries have accepted the format as the 3G system of choice. That WCDMA is the natural 3G progression of GSM should come as no surprise, as GSM has the dominant market share worldwide.

Like all initial deployments of a new technology, there were and are early technical issues which need to be, and will be, solved in the first deployments in Japan and Europe. This may lead to subtle improvements in the standard. For the most part,

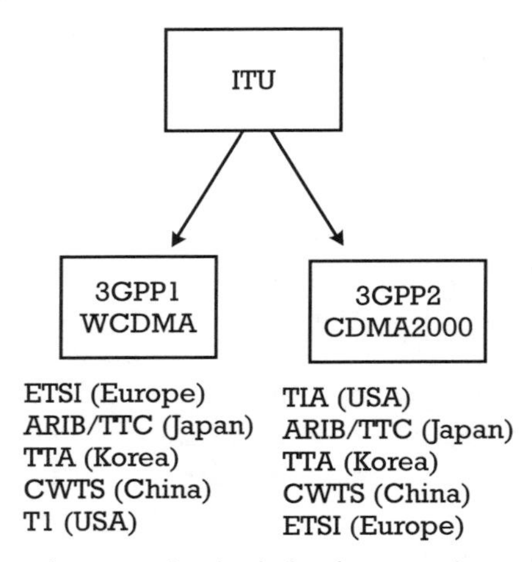

Figure 9.1 The ITU serves as the governing body for the two primary 3G standardization groups. Shown are the main standard bodies that lead and approve the process and standard in each country.

however, the system is ready to go as the capital and spectrum needed for Greenfield deployments becomes available.

9.3 Features and Benefits of WCDMA

Most “noninsider” followers of the wireless industry most likely think of the wonderful multimedia capabilities that 3G services hope to offer. Downloading the latest photo of their children while driving down the highway away on business, having a video conference while on the train, or playing a multiplayer video game during a school break—these are the images most would have of what 3G could offer. The reality is many engineers and executives are looking forward to 3G technology because of the added voice capacity and spectrum efficiency it can bring, with the high-speed data serving as another revenue booster along the way. It is expected that the 5-MHz channel in WCDMA will be able to support more than 100 simultaneous voice calls. And because WCDMA uses CDMA, there are no frequency reuse issues to deal with.

Earlier, it was mentioned that WCDMA was developed without any constraints from needing to be backward compatible, as it would deployed from scratch. This is not altogether correct. While the air interface is entirely new, the format was designed to operate on the same core network as GSM, GSM-MAP. This means WCDMA *radio network subsystems* (RNSs), the UMTS name for the BSS, can be deployed right alongside GSM BSSs and use the same core network, which includes the MSC, *equipment identity register* (EIR), HLR, VLR, and AC. In addition, as most networks will use GPRS systems, support for GPRS data is compatible with the WCDMA system via the use of GPRS support nodes. Thus, even though different RF subsystems will be needed to support 2G, 2.5G, and 3G formats in the GSM-to-WCDMA transition, the core network can remain the same and thus offer multimode services to a common area (see Figure 9.2).

WCDMA, when compared to the GSM (TDMA) system it will complement and eventually replace, will utilize many new techniques to allow for the added voice capacity and high-speed data rates. Most of these new (in reference to GSM) techniques are already used in the CDMA2000 system. They include soft handoffs (in WCDMA, as in GSM, these will be referred to as handovers), Rake receiving, fast power control, advanced FEC such as convolutional encoding and turbo coding, and new modulation and spreading of the physical channels.

9.4 TDD and FDD Modes

The system is designed to operate in two modes, FDD or TDD. You will remember from earlier chapters that this pertains to whether the uplink or downlink share the

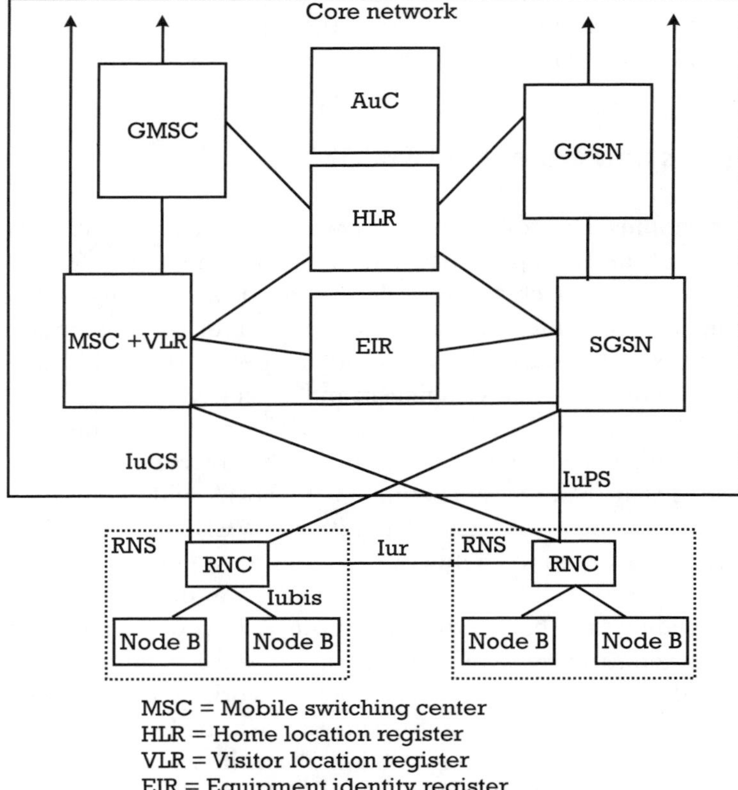

Figure 9.2 The layout of the core network in WCDMA networks is very similar to GSM networks. Note that the interface names are different.

same frequency and share transmit time or they are given separate frequency bands (see Figure 9.3). TDD is very convenient for locations where spectrum is not available because in FDD mode, 5 MHz would be required for both the uplink and downlink for each WCDMA channel, separated by a set amount of spectrum.

In TDD mode (see Figure 9.4), the 5-MHz channel is divided into 10-ms frames; these frames are then divided into 15 time slots of 666 μs each. These time slots can then be assigned for transmit or receive at the base station in order to have the uplink and downlink share the same frequency. This allows for a second benefit, the ability to make the data transmissions asynchronous (see Figure 9.5) (i.e., more slots can be

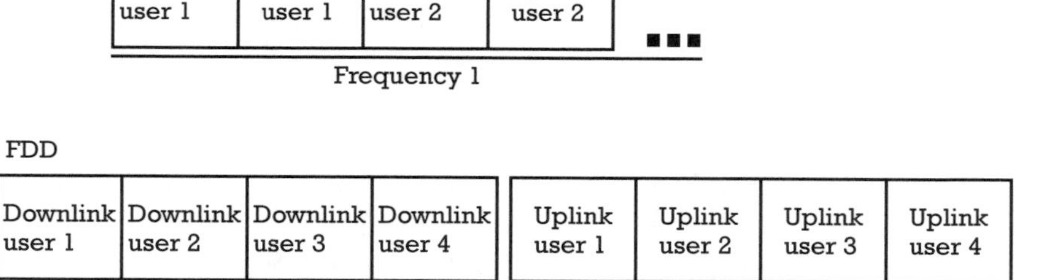

Figure 9.3 In TDD mode, the uplink and downlink share the same frequency; in the FDD mode, the uplink uses a separate frequency band than the downlink.

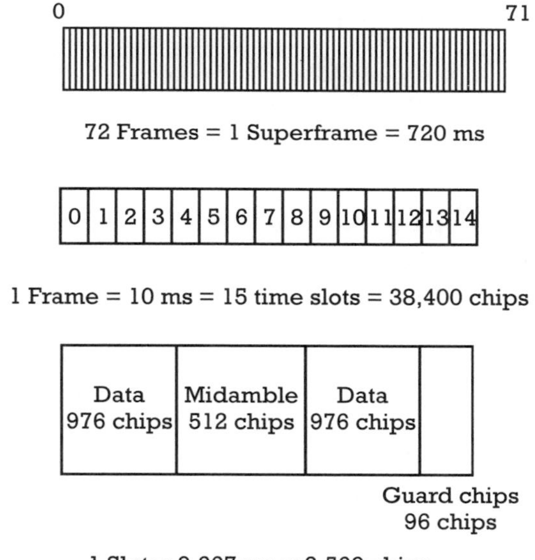

Figure 9.4 The TDD data frame structure is very similar to the FDD frame structure, although each time slot can be for uplink or downlink.

allocated to the downlink than the uplink, so that large amounts of data can be sent to the mobile more efficiently if the mobile does not need to send large amount of data backup). This is the model of the current Internet world, where the downstream data use is significantly higher than the upstream.

A third benefit of using TDD is that it allows for better power control (see Figure 9.6). This is because both the uplink and the downlink share the same frequency, so the fast-fading characteristics will be similar in both directions. Thus, a transmitter can predict the fading based on the received signal and adjust

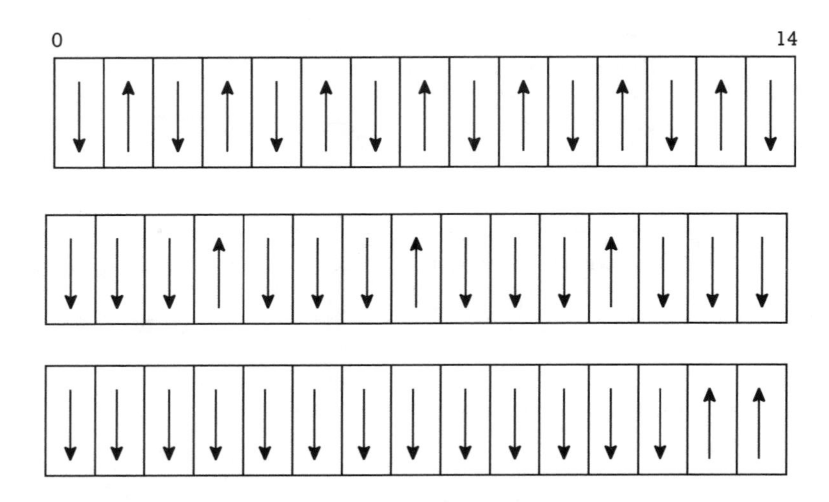

Figure 9.5 One of the advantages of the TDD mode is the flexibility to cater to asynchronous data traffic. If the data traffic is more downlink intensive, as many predict, more downlink channels can be added per frame.

accordingly, making the closed-loop power control system unnecessary. The problem with this, of course, is the basis for the power control decision is based on received power; thus, the interference must still be reported, as interference will limit quality. In addition, because the quality of both links can be determined by the uplink, the network can make better decisions on from which base station would be best to transmit the downlink signal to individual mobiles.

There are reasons why most operators will not be deploying the TDD version, however—or at least deploying it only in "hot spots." This is primarily due to interference from what is called *TDD power pulsing*. That is, because the mobile must pulse its power, it will generate audible interference. As the mobile increases its speed, these pulses will need to be shorter and will increase this interference.

Perhaps even more importantly, the TDD system is very prone to interference from other base stations. Because the same frequency is used throughout the network, it is possible that a base station may be transmitting to a mobile on the same slot as another mobile trying to receive from a neighboring base station. The end result would be the second mobile would not be able to receive its desired signal. This can be alleviated if all of the base stations were synchronized (as in the TD-SCDMA system) and if the asymmetry of the time slots were set constant (i.e., the same amount of downlink and uplink time slots was used across the network). Both of these are not in place in the WCDMA system; thus, the system will probably be used more in limited areas to complement FDD operation. It is thought that TDD will be used extensively in indoor applications, where interference would be limited but high-speed downlink rates would be desired.

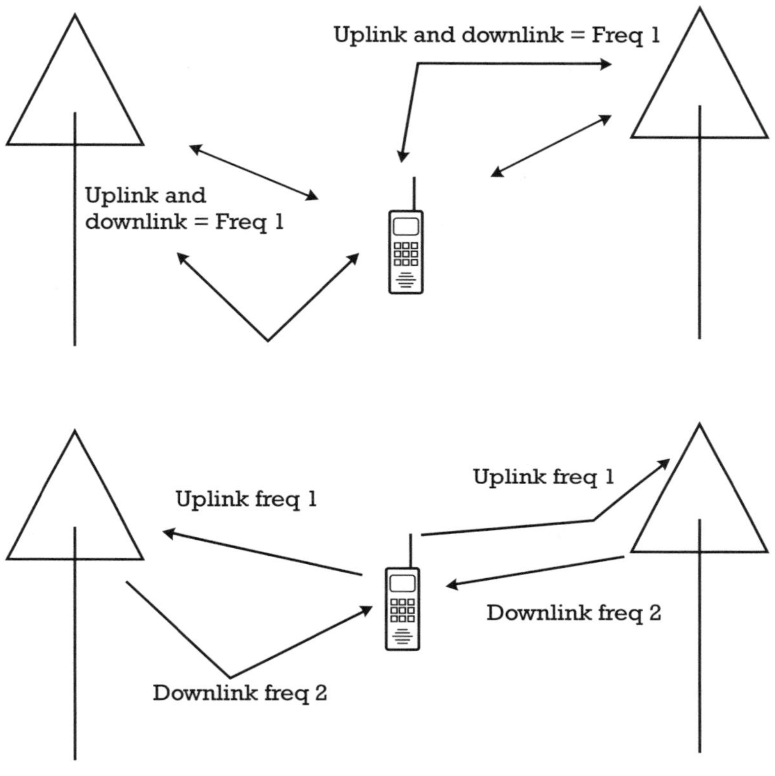

Figure 9.6 Another advantage of TDD mode is that power control can be more efficient, as the uplink and downlink use the same frequency and, hence, have the same fade characteristics. In FDD mode, the uplink and downlink will both fade differently and hence need independent power control.

As the physical channels of the format are discussed, it is important to understand that in FDD mode, channels are defined by their code and frequency; in TDD mode, they are defined additionally by the associated time slot. For the purposes of this text, unless otherwise stated, the material will pertain to the FDD mode—but, again, the functionality is similar.

Although TD-SCDMA, discussed briefly in the last section, is looked at as an entirely different standard, it is almost identical to the TDD version of WCDMA, except it uses a narrower bandwidth and synchronous operation.

9.5 Introduction to Nomenclature

Several of the nomenclatures used in 2G systems have been changed in the WCDMA standard (see Figure 9.7). The mobile station is referred to as the *user equipment*

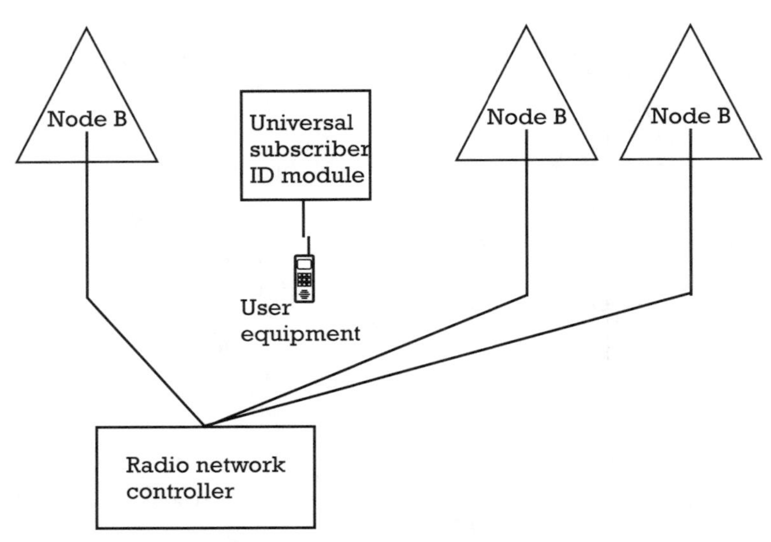

Figure 9.7 In WCDMA, the nomenclature for the components of the RNS, as shown, are different.

(UE). The BTS is called the node B. The BSC is the *radio network controller* (RNC). The subscriber identity module, or SIM card, is now the *universal subscriber identity module* (USIM) card. Finally, as already discussed, the BSS is called the RNS.

There are several other terms that a technician or engineer might hear used interchangeably when discussing WCDMA that can cause a newcomer a lot of confusion. The full name of the WCDMA standard is the UMTS terrestrial radio interface, in either TDD or FDD modes. This is thus referred to as UTRA-FDD or UTRA-TDD. Quite often, the UTRA network might be referred to as the UTRAN. In addition, the ITU calls this same standard IMT DS.

9.6 Slot and Frame Structure

While WCDMA obviously uses CDMA techniques to place multiple channels in the same frequency, identifying those channels with the codes discussed in this chapter, it also uses the time domain via its slot and frame structure (see Figures 9.8 and 9.9).

The channel is divided up into 10-ms frames, each frame having 15 time slots that are 666 μs. Within the time slots are fields that contain either the user data or the control messages. The actual configuration will vary depending on the logical and physical channels. However, because the chip rate in WCDMA is 3.84 Mcps, each time slot is made up of 2,560 chips—meaning with no protection or spreading, one time slot could conceivably transmit 2,560 bits. However, because the spreading factors range from 4 to 256 for the uplink, and 4 to 512 for the downlink, the bit

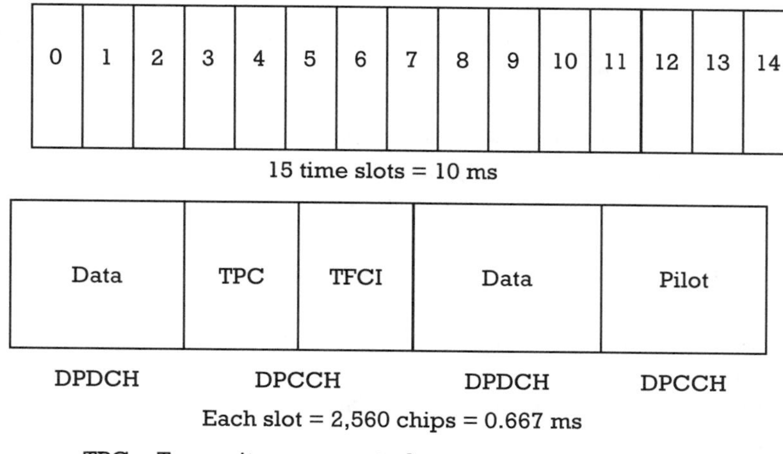

Figure 9.8 The downlink slot structure for data channels. Note that the control channels and data channels are combined in the same slot.

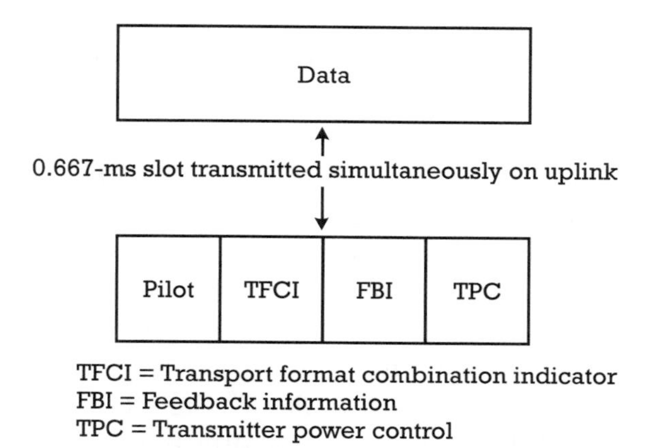

Figure 9.9 The uplink slot structure is slightly different. The uplink transmits control and data channels on the I and Q channel simultaneously in order to avoid bursty transmissions.

rate will range from 7,500 bps to 960 Kbps. In TDD mode, the spreading factors are from 1 to 16, so the corresponding bit rates per slot are 240 Kbps to 3.84 Mbps.

9.7 Logical and Transport Channels

In order to understand how the information is delivered in the uplink and downlink, the three channel concepts in the UTRAN must be understood. These three general types of channels include logical, transport, and physical. The logical channels define how and with which characteristics the data will be transferred by the physical channel. Transport channels, which (along with the logical channels) are also called the layer 2 channels, define how the data will be transferred by the physical layer. Physical channels, also called the layer 1 channels, define the exact physical characteristics of the radio signal. Unfortunately, unlike the CDMA2000 standard, there is no differentiation in the nomenclature of the channel abbreviations, so it can cause a lot of confusion.

Thus, you have data in a logical channel that is then processed into a transport channel. From there, the transport channel maps to a physical channel. As an example, the *broadcast common control channel* (BCCH), which is a logical channel, is mapped to the transport channel known as the *broadcast channel* (BCH), which is then transmitted on the *primary common control physical channel* (P-CCPCH) (see Figure 9.10).

There are seven logical channels defined. These include:

- *Paging control channel (PCCH):* The PCCH is used in the downlink only. It is used to send paging information and various notifications to mobiles.
- *Broadcast control channel (BCCH):* The BCCH is used in the downlink only. It broadcasts system information specific to the cell.
- *Dedicated control channel (DCCH):* The DCCH is a bidirectional channel. It transfers dedicated control information.
- *Common control channel (CCCH):* The CCCH is a bidirectional channel that transfers control information.
- *Shared channel control channel (SHCCH):* The SHCCH is bidirectional. It is only used in the TDD mode to transfer shared channel control information.
- *Dedicated traffic channel (DTCH):* The DTCH is the bidirectional channel that carries user information.
- *Common traffic channel (CTCH):* The CTCH is a downlink-only channel used to transfer dedicated user information for a group of users.

The transport channels define how the data is to be transferred to the physical channel. It thus makes sense that all of the transport channels are unidirectional. Like the logical channels, the transport channels contain dedicated and common channels. The common transport channels are as follows:

- *RACH:* The RACH is an uplink channel that is used for nonreal-time control or traffic data.
- *Common packet channel (CPCH):* The CPCH is an uplink channel used for bursty data traffic in FDD mode.
- *Forward access channel (FACH):* The FACH is a downlink channel which carries user data.

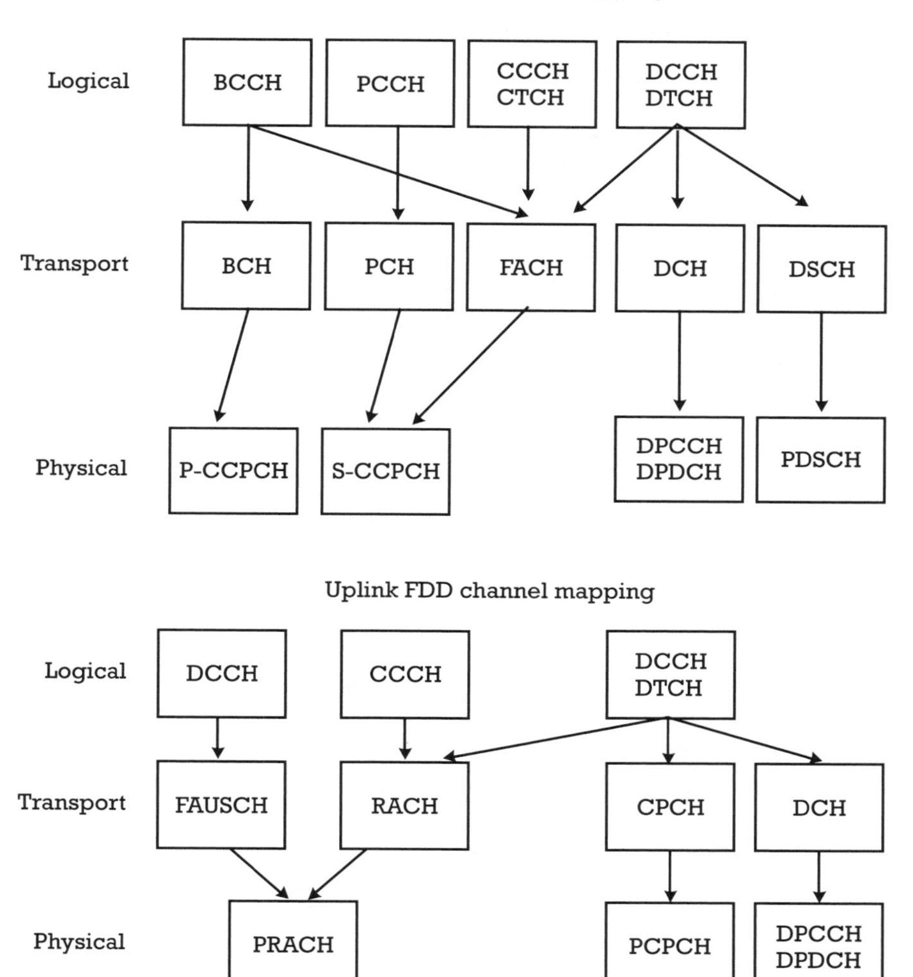

Figure 9.10 Logical channels map to transport channels, which map to physical channels. Note that not all the channels are shown—many of the physical channels do not have transport/logical channels associated with them.

- *Downlink shared channel (DSCH):* The DSCH is a downlink channel that carries dedicated control or traffic data and is used in conjunction with a dedicated channel (DCH).
- *Uplink shared channel (USCH):* The USCH is an uplink channel for the TDD mode of operation, and it contains dedicated control or traffic data.
- *BCH:* The BCH is a downlink channel that broadcasts system and cell-specific information. It is specific to the node B as it provides the UE with system

information, neighbor radio channels, and configurations. It also gives the UE the available RACHs and the associated scrambling codes.

- *PCH:* The PCH is a downlink channel used for transmission of paging and notification messages, and it is associated with the physical channel called the *page indication channel* (PICH).
- *DCH:* Each UE will have one DCH assigned, which will be used to transfer UE-specific data.
- *Fast uplink signaling channel (FAUSCH):* A dedicated uplink channel used for sending UE specific signaling data.

9.8 Physical Channels

Once the logical channels are mapped to the transport channels, they can then be mapped to the physical channels, which is how they are transmitted out from the Node B and the UE. You will recall that there are two modes of operation, FDD and TDD. The makeup of the physical channels is different for the different modes of operation.

The physical channels include:

- *Common pilot channel (CPICH):* This is a downlink channel that serves as the phase reference for the downlink channels. It has a fixed rate of 30 Kbps. There are two types of CPICH, a *primary* (P-CPICH) and *secondary* (S-CPICH). The primary serves as the reference in order to decode the other channels. The CPICH is an all-ones channels, just as in 1XRTT, scrambled with the code of the node B. Once the mobile obtains frame and slot synchronization (via the SCH process described later), it can use the pilot to have an absolute coherent reference. The secondary pilot is used in smart antenna or narrow antenna beamwidth applications.
- *Primary common control physical channel (PCCPCH):* This is a downlink channel used to transmit the BCH, which contains system and cell-specific information. It has a fixed rate of 30 Kbps.
- *Secondary common control physical channel (SCCPCH):* This is a downlink channel that carries the FACH and the PACH for mobiles that have not registered yet. While there must be at least one SCCPCH per carrier, more can be added depending on the number of UEs that are allowed to operate in the system. The channel transmits only when data needs to be sent.
- *SCH:* This is a downlink channel that contains no information but is used for synchronization and cell searching by the mobile. There are two subchannels, the primary and secondary. The primary used a 256-chip sequence that is identical in all the cells and allows the mobile to gain initial slot synchronization.

The secondary synchronization channel is used to gain complete slot and frame synchronization. This process is explained in detail later in this chapter.

- *Physical downlink shared channel (PDSCH):* This downlink channel carries the DSCH. It is always associated with a downlink DCH.
- *Acquisition indicator channel (AICH):* This downlink channel is used for the random access procedure. It carries *acquisition indicators* (AIs), which is essentially the green light to a UE to proceed with a transmission on the access request. Like 1XRTT, the mobile will attempt to access the node B at a low level in a specific reverse access channel via the PRACH. The node B then sends an AI on the AICH as to whether the mobile needs to proceed with the process or stop the process. If the node B did not get the message, the UE will increase its power until it gets the green light. This process is known as *dynamic persistence.*
- *PICH:* This downlink channel tells the UEs that they have a page message waiting for them on the PCH. It is always associated with a specific PCH, as the channels tells a specific set of UEs whether they should look at the next frame of a specific S-CCPCH for the PCH message.
- *Access preamble acquisition indicator channel (AP-AICH):* This downlink channel is used with an associated CPCH to facilitate sending bursty data traffic on the shared CPCH uplink logical channel.
- *CPCH status indicator channel (CSICH):* This downlink channel carries the status of the CPCH, again used for bursty data traffic and works similar to the PICH.
- *Collision detection/channel assignment indicator channel (CD/CA-ICH):* A downlink channel used to tell the UE when the channel assignment is active or inactive.
- *Dedicated physical data channel (DPDCH):* This is a dedicated uplink or downlink channel used for dedicated data. It carries the DCH.
- *Dedicated physical control channel (DPCCH):* This is an uplink channel that carries control information.
- *PRACH:* This uplink channel carries the RACH.
- *Physical common packet channel (PCPCH):* This uplink channel carries the common packet channel.

9.9 Spreading

WCDMA, like IS-95, uses a complex radio waveform that is encoded to ensure only the desired recipient is able to correlate the sequences. Each physical channel is spread with unique variable spreading sequence. This spreading process is explained more thoroughly in the IS-95 and 1XRTT chapters of this text. In WCDMA, the downlink physical channel is spread by a combination of a channelization code and a scrambling code.

The channelization code is used to increase the data rate to the transmit rate. At its simplest level, this channelization process transforms each data symbol into several chips. The ratio of symbol:chips is called the *spreading factor*.

On the downlink, the transmitted symbol rate of WCDMA is 3.84 Mbps and the modulation is QPSK, which, you may recall from earlier chapters, means that each transmitted symbol represents 2 bits of information. This means that the rate at the modulator is essentially 7.68 Mbps (3.84×2). Thus, if the data rate (i.e., the actual data we are trying to transmit) is at 15 Kbps, it will need to be spread by a factor of 512 in order to create the required chip rate for transmission (see Figure 9.11). Likewise, if the actual data rate is 1,920 Kbps, it will only need to be spread by a factor of 4 to get the same chip rate. Realize that this means the processing gain is not as high; thus, for the same reception performance and error protection, the lower spreading

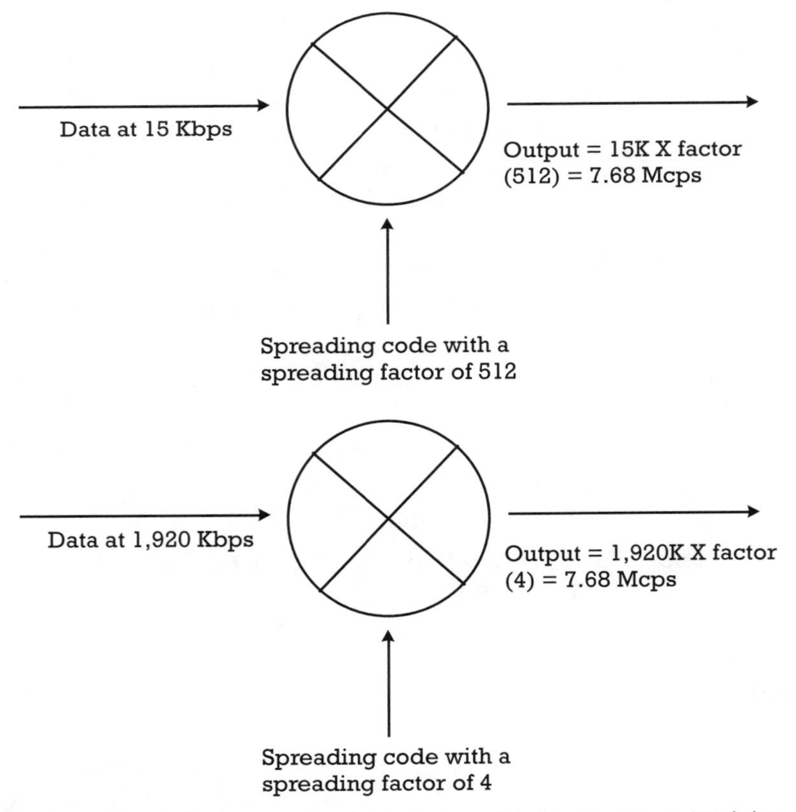

Figure 9.11 Spreading codes increase the data rate to match it to the transmitted data rate. The spreading factor changes to match the actual data rate that needs to be transmitted.

factor will have more errors than the higher spreading factors (i.e., it will require more power or better sensitivity to transmit higher data rates).

The channelization process uses orthogonal codes, specifically OVSF, to spread the data. Variable orthogonal codes are discussed at length in the 1XRTT chapter of this text, particularly see Figure 9.4. In order for OVSFs to be used to separate signals, they must be synchronous to remain orthogonal. Within a particular cell's range, this is the case; thus, the OVSFs are used to separate users in the downlink. In the uplink, however, the UEs are not synchronized, nor are the node Bs in the downlink (except in the TDD mode with uplink synchronization). This means that the OVSF channelization is not sufficient for separating UEs on the uplink, nor for separating node Bs on the downlink, as without synchronization, the signals could potentially interfere with each other. Thus, a second set of codes is used to spread, in a process called *scrambling*. Thus, first in the chain of the transmitters is the channelization process using the OVSFs, followed by the scrambling process, which will identify the node Bs as well as the individual UEs.

The scrambling process uses PN codes, similar to those used in IS-95. After being spread by the OVSF to get the chip rate required, the bits are then spread again by a PN code (see Figure 9.12). On the uplink, millions of different PN codes can be used to identify the UEs, as one long code (or Gold code) can be masked with a unique UE code (e.g., a serial number) to create a unique sequence. Thus, on the uplink, there should never be an issue with not enough PN sequences. The downlink, just as in IS-95, uses a short code with 512 different variations. Thus, each node B will be assigned one of the 512 codes (in IS-95, these were called PN offsets). These 512 primary scrambling codes are grouped into 64 code groups, each with

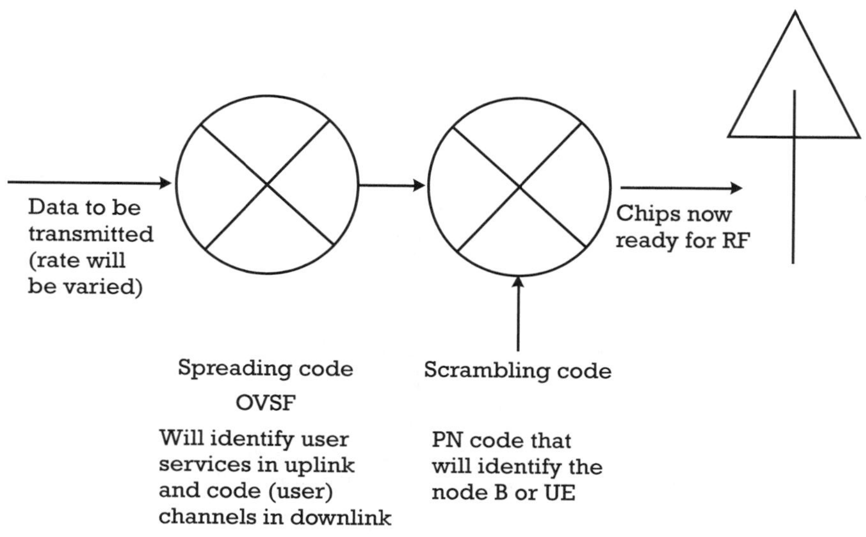

Figure 9.12 The spreading and scrambling process.

eight codes. This speeds up the synchronization process, as you will see in the next section.

9.10 Synchronization

You will remember from IS-95 that when powered on, an IS-95 mobile would immediately begin searching for the pilot channel in order to get synchronized to the base station and the network. The pilot in IS-95 did not have any information but was simply a PN code with a set data rate that could easily be correlated to by the mobile to begin reading the other channels. In WCDMA, you have already seen that the pilot plays a slightly different role.

Synchronization in WCDMA comes from the *primary synchronization channel* (P-SCH) and the *secondary synchronization channel* (S-SCH) (see Figure 9.13). These channels are not spread using the OSVFs or the PN codes. Instead, they are spread by synchronization codes. There are two types of synchronization codes, primary and secondary—logically, the primary code is used on the P-SCH, and the secondary code is used on the S-SCH. The primary code is the same in all of the cells, and it contains a 256-chip sequence that is transmitted only during the first 256 chips of each time slot. This enables the UE to gain slot synchronization with the base station. At this point, a UE knows the start and stop of each time slot, but it

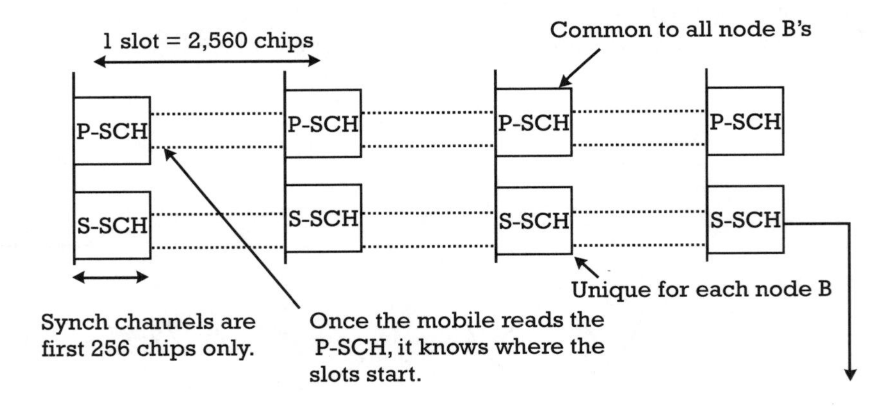

Each S-SCH can be one of 16 SSC codes. Fifteen SSC codes form a SSC sequence. There are 64 SSC sequences. Once the mobile reads 15 codes, it knows the sequence (code group). Once it knows the code group, it knows where the frame starts/ends. Once it knows the code group, it has narrowed down the scrambling code to 1 of 8 possibilities. It can then correlate each scrambling code candidate with the CPICH of the cell until it finds the correct one. Once it has the scrambling code, it can read the BCH and continue with call processing.

Figure 9.13 Synchronization in WCDMA uses the P-SCHs and S-SCHs.

does not know which time slot it is on nor where the frame boundary is. There are 15 time slots in each frame. This is where the S-SCH comes in.

There are 16 different secondary synchronization codes. One of these codes is sent on the first 256 chips of each time slot; thus, one secondary SCH sequence, which relate to a scrambling code group, consists of 15 synchronization codes. There are 64 different sequences and, thus, 64 scrambling code groups. These sequences are arranged in such a way that after reading the entire sequence, the UE can identify the scrambling code group. Once it identifies the code group, it is able to determine before which synchronization code the frame begins, and, hence, it has achieved frame and slot synchronization.

These scrambling code groups also point to which scrambling code is being used to identify the base station. You will recall that the scrambling codes were divided into 64 code groups, each with eight codes. Thus, after getting frame synchronization by identifying the scrambling code group, the UE has also narrowed its choices for a scrambling code to eight and can simply try and decode the CPICH with each of the eight. Once it gets the right scrambling code, it can then read the BCH information to get better timing reference in order to begin monitoring the P-CCPCH.

9.11 Power Control

Like all CDMA-based systems, WCDMA has fast and efficient power control implemented. It is vitally important on the uplink that all of the UE's signals reach the base station antenna at the same levels, or they will face problems of interference from near-far issues (explained in depth in earlier chapters). The mobiles further away from the base station need to be transmitting at higher power levels than those close to the base stations. In the downlink, good power control is also important, especially in WCDMA's implementation. Because other base station signals are not orthogonal to each other, but use the same frequency, the signals will interfere with each other. Hence, it is vital that the base stations transmit at as low a power level as possible.

WCDMA, like IS-95 and 1XRTT, uses an open loop and a closed loop in its power control technique. The open loop is used during initial access. It is simply a measurement of the received signal in order to give an estimate of the path loss. The problem with this method is that the uplink and downlink are on different frequencies; hence, the fading characteristics can be substantially different. In order to compensate for this, the closed loop is used.

The closed loop involves making a measurement of the received signal in every time slot (667 ms), after which a power control bit is sent to either step up or down the transmitter being controlled. This is done on the uplink as well as the downlink. Like IS-95, there is no neutral power control bit—the mobile or base station will either step up or down every time slot (every 667 ms). Note that the determining factor for whether to step up or down is a power measurement, not a FER measurement. This is done as a power measurement because it is much faster, as it does not require demodulation of the signal. The system will try and manipulate the transmitter to hit a predetermined *signal-to-interference ratio* (SIR) target at the receiver. This, however, can cause some inaccuracies in the power control, because different coding and interleaving methods and parameters can mean that the SIR target is not appropriate for the quality needed or desired for a certain bit stream. Hence, this SIR target needs to be dynamic in nature. This is where what is called the inner and outer loop come into play.

The inner loop is essentially the technique described in the last paragraph, where a received signal is measured and a power control bit, either to step up or down, is punched into the time slot to maintain the SIR target level. The outer loop involves decoding the received signal and performing a quality measurement (FER) in order to decide whether the current SIR target is adequate or can be lowered or increased.

In the case of soft handovers in WCDMA, there are several differences from IS-95. You will remember in IS-95 that during a handoff, the mobile would always follow the power control from the base station telling it to go down in power, if conflicting power control bits were received. In WCDMA, this is not a hard and fast rule, and a special algorithm is used to make this determination. In addition, WCDMA implements a technique called *site-selection diversity transmission* (SSDT). Essentially, this is a method of power control during soft handover, where only one base station is selected to transmit the power control bits (see Figure 9.14).

In the downlink, several options are available in deployment. While fast closed-loop power control offers the most effective power control, the network can simply use a slow closed-loop power control method for several of its channels. Similar to IS-95, this involves making FER measurements prior to sending a request to increase or decrease power.

9.12 Handovers in WCDMA

Like IS-95, WCDMA uses soft handoffs for digital-to-digital handoffs from cell to cell. In a soft handover, the UE is able to receive and combine the signal from two or more base station simultaneously. Like IS-95, the receivers use a Rake receiver. You will remember from earlier chapters that the Rake receiver can be used to receive and combine multipath signals in addition to signals from different base stations. While multipath signals are simply time-delayed versions of the same signal from the same transmitter (hence, time-delayed versions of the same spreading code), each base station will have a truly different spreading code.

Softer handoffs are also used in WCDMA. Here, two signals from different sectors of a base station are combined. This looks just like a normal soft handoff to the

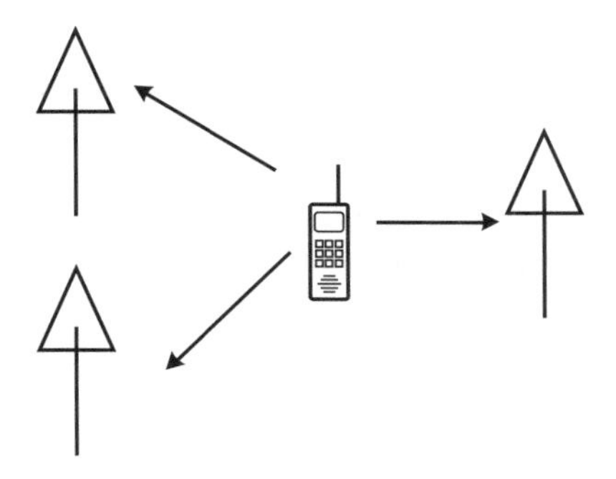

In the uplink, the mobile is received by
all three handover node B's.

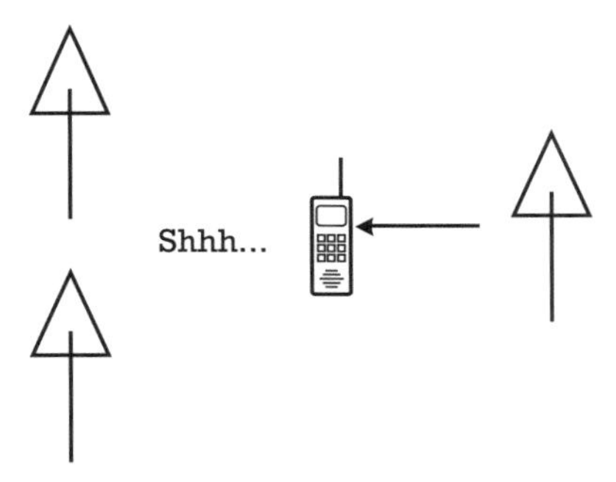

In the downlink, only the best node B transmits;
the others lower their signals.
Shhh...

Figure 9.14 SSDT allows the effect on the RF during soft handovers to be lessened by lowering the power of some of the base stations in handoff with the mobile.

UE, but to the network it is different, as the base station is able to combine the receive signals without having to use the network's transport resources.

Just as in IS-95, WCDMA uses add and drop threshold to decide whether base stations are candidates for soft handoff. All of the decisions are actually made by the network, so measurements of the received signal at the UE and the node B are reported to the RNC. The different node Bs are classified into one of three sets for each UE in the network: the active, monitored, or detected. Active node Bs are those where the reported signal strength exceeds the addition threshold. This addition threshold, which is similar to t-add in IS-95, is dynamic in WCDMA, as it is in 1XRTT, so that the network can decide the correct level for the desired level of performance. This is the same for the drop threshold, which is similar to t-drop in IS-95/1XRTT.

The monitored set is similar to the neighbor list in IS-95. The set of nearby cells have been identified as potential actives by the network but are not yet active. This list is given to the UE by the network.

The detected set simply includes all of the other cells in the entire network. The UE will monitor these, though less often than the monitored set, as the neighbor list might not be totally accurate.

As briefly mentioned in the power control section, WCDMA can use a different power control method when in soft handover called SSDT. In normal handoffs, the UE will be communicating with multiple node Bs. This is not always most efficient, as every downlink channel will add to the overall interference in the network. Using this system, the UE can select the node B from which it has the best reception as the primary cell. Naturally, this is all very dynamic. This primary cell will continue to transmit normally. The other node Bs involved in the soft handover, however, will lower the power in the DPDCH addressed to the UE. This reduces the amount of interference in the overall network and lessens the impact of high soft handoff factors. It is important to note that it is the UE that selects the primary cell out of the active list to be the primary. This has an advantage of speed, as primary cells can be selected quickly without the need for contacting the network.

9.13 Modulation

WCDMA uses different modulation techniques on the uplink and downlink. The modulation rate is 3.84 Mcps, although it originally was 4.096. The compromise was made to make the rate close to CDMA2000 rates.

On the uplink, WCDMA uses dual-channel QPSK. This means that different data streams are sent on the I and Q channels (see Figure 9.15). In this case, the DPDCH (dedicated user data, either voice or data) is sent in one channel, and the

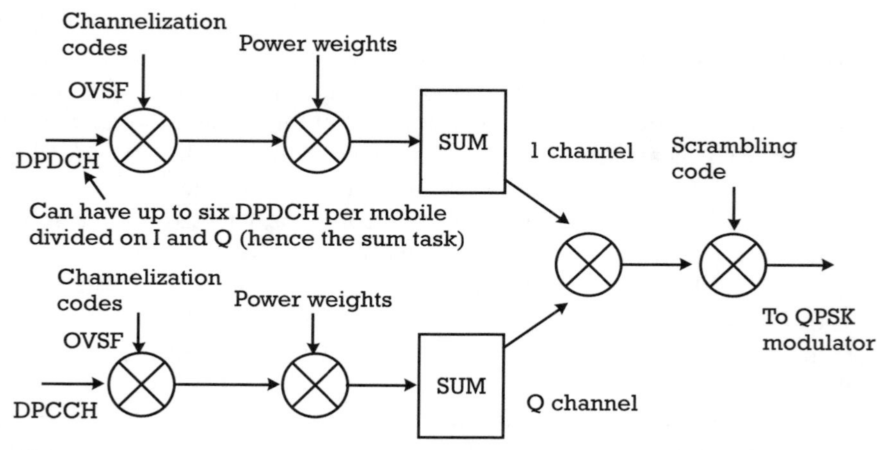

Figure 9.15 In the uplink, multiple channels are sent in the I and Q channel, each having its own power control weight.

DPCCH (dedicated control information) is sent in the second channel. While the control channel is continuously transmitted, the user data uses discontinuous transmissions. By using this dual-channel modulation, interference caused by the transmitter turning on and off is avoided. You might recall the electromagnetic interference from discontinuous transmissions was a serious problem with GSM phones due to this issue. In addition, as the uplink transmission is the largest source of battery drain, the modulation scheme uses very complex scrambling methods, similar to CDMA2000 (see Chapter 8 for an in-depth discussion).

In this complex process, the control channels are sent down the Q path, while the first data channel is sent down the I path, with additional data channels getting split between I and Q paths. Again, you should read through the more in-depth section in Chapter 8 for the method, which then involves power weighting and Walsh code rotations.

The downlink of WCDMA uses QPSK modulation. The downlink does not have the concern of battery life or power amplifier complexity that the uplink does. The two dedicated physical channels, the DPCCH (control channel) and the DPDCH (data channel) are time multiplexed in the downlink. As discussed earlier, in the uplink this caused problems because it would result in a pulsed transmission, as the DPDCH is a discontinuous transmission channel. The downlink, however, also transmits the host of common channels, all of which are transmitted continuously. Hence, there are no pulsing issues that need to be resolved via complex modulation in the downlink.

9.14 Multiple Channel Codes

In order to give a user the highest transmission data rates, the WCDMA system can combine multiple channel codes by coordinating the data sent on the multiple physical channels. Control messages are used to instruct how the multiple channels are combined. In addition, two indicators are used: the *transport format indicator* (TFI) and *transport format combination indicator* (TFCI) codes. The TFCI is sent with the physical channels to give information on the format of the channels. It is decoded and used to create the TFI, which is the transport channel version.

9.15 Conclusion

While CDMA uses significantly different nomenclature than its earlier CDMA relative, CDMA2000, it is quite similar in many ways. The similarities begin with the use of CDMA techniques but go much further, including similar power control, modulation, spreading, and scrambling techniques. There are significant differences, however. WCDMA uses a different slot structure and has a very different method for obtaining synchronization between the UE and node B. It has become clear, at the time of writing this text, that although delayed, WCDMA will be one of the three main voice formats for the foreseeable future, along with CDMA2000 and GSM.

CHAPTER 10

The Basics of Field Test

10.1 Introduction

In order to perform any job correctly, a person must have the best tools available to them. More importantly, a person must know how to effectively choose and use these tools. Many carriers have implemented a "cookbook" approach to network maintenance (i.e., having technicians follow a completely structured test procedure without really understanding what it is they are testing—or often, more importantly, what they are *not* testing—nor what the instrument they are using really tells them). While this approach can work for a time, the most efficient networks and service shops are those maintained by technicians who have a good grasp on the technology, understand the "whys" in addition to the "hows," and can make insightful decisions on improving the performance of a system, be it a base station or a mobile station.

This chapter and Chapters 11 through 14 are designed to give a technician and technical manager a basic understanding of many of the tools they have available to them as well as provide an overview of the primary tests that are performed in the field for each technology discussed. It is not designed to be a test engineering reference, nor will it review tests that are more research and design oriented. Additionally, this book is designed as an overview of general procedures and performance, and will not get specific in setups or front panel operation. While standards dictate many of these test procedures along with pass/fail specifications, many carriers, service shops, and even manufacturers will implement their own tighter or even occasionally looser specifications as well as their own setup procedures for certain tests. This is another reason this text will stay somewhat generic in describing the test procedures and will not list specific pass/fail specifications.

10.2 Types of Cellular and PCS Technicians

Wireless technicians can be divided into three groups, for the purposes of this text:

- *Base station technicians:* These technicians are tasked to commission (install) base stations and ensure proper operation. Typically, this involves testing the base station transmitter and receiver, as well as testing the antenna and associated cables. Base station technicians usually work closely with field optimization and performance engineers as the task of optimizing network performance and maintaining and installing base stations is closely related. Base station technicians are also sometimes tasked with maintaining the backhaul (i.e., the connection between the base stations and the base station controller, which might be a T1 or microwave link). Base station technicians will often use instruments such as spectrum analyzers, power meters, frequency counters, modulation analyzers, multimeters, scalar network analyzers, and signal generators. Very often, a tool called a *service monitor* or *communication test set* is given to a technician team. A communication test set incorporates many of the instruments mentioned earlier into one box, saving the technician from carrying multiple instruments to often hard-to-reach cell sites. This leads to a very important consideration in test equipment for the base station technician: portability. Of course, a down cell site could mean losses of thousands of dollars for a carrier, so speed is equally important.
- *Network technicians:* The network technician is tasked with maintaining the operation of the switch. This generally involves ensuring the switch hardware and connections are working properly, performing some protocol analysis, and monitoring the network for alarms, working closely with base station technicians. The network technician's background is generally more telephony-based and information system–based, but a good understanding of the air interface is very helpful. Network technicians may use such test instruments as protocol analyzers, which decode much of the information sent on the interfaces, and BER testers, which can verify that a transmission path is good.
- *Mobile station technicians:* The mobile station technician generally works in a shop, either operated by the wireless carrier or an independent repair or sales organization. Many technicians will interact directly with the customer as the customer brings in a phone for repair. A good understanding of the technology can really help such a technician quickly determine if the customer is the victim of poor network performance or truly has a malfunctioning mobile station. Eliminating the "no fault found" repair orders has become an important goal of carriers, as they often must pay a fee for every phone sent to a higher level service center even if there turns out to be no problem with the phone. The mobile station technician also uses a variety of test instruments, such as power meters, frequency counters, multimeters, spectrum analyzers, signal generators, and modulation meters. Additionally, testing mobile stations often requires a base station simulator in order to test call processing and simulate a

call environment to perform certain tests. Again, a service monitor or communication test set is used, which includes many of these instruments. Because repair shops deal with repairing low-cost mobile phones, and phones not in service are phones not making money for a carrier, turnaround time, throughput, and cost are often the primary consideration in test equipment. This relates to equipment that is very easy to use, tests quickly, and is inexpensive—a tall order, indeed.

10.3 Introduction to the Radio Transceiver

The basic sections of a two-way radio are essentially the same, regardless of the technology. Perhaps the term radio is the most important word in the previous sentence. Many people forget that their cellular phone is really just a two-way radio, not tremendously much departed from the old walkie-talkie one might have used as a child (well, that might be a stretch...).

Essentially, any cellular radio, be it in the form of the mobile station or the base station, can be broken down into nine sections (although, naturally, base stations will have more hardware for interconnecting to the bigger picture) (see Figure 10.1):

1. *Transmitter:* This generally consists of the PA as well as much of the power control hardware.
2. *Receiver:* This is where preamplification, filtering, and down converting to a workable IF takes place.

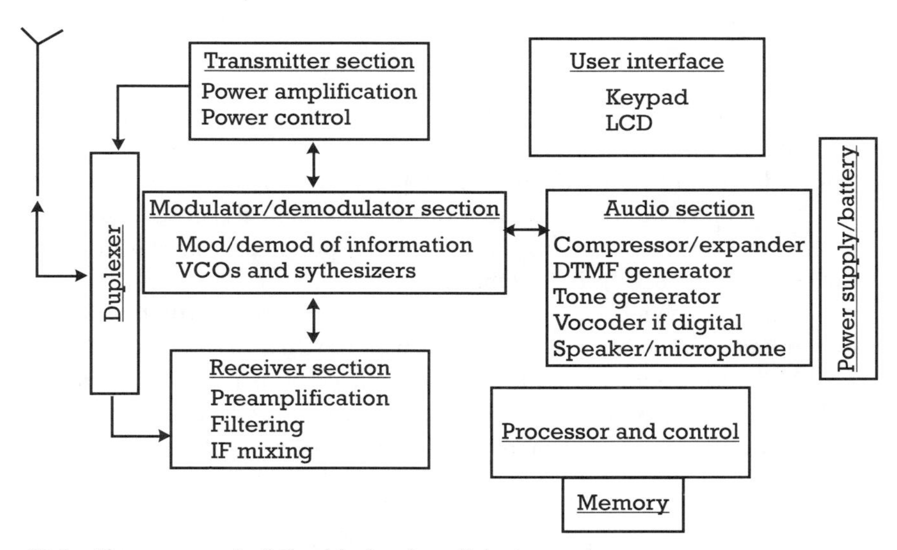

Figure 10.1 The common building blocks of a cellular transceiver.

3. *Modulator/demodulator:* This is where the audio is modulated on the carrier for transmission and demodulated back to analog audio for reception. It also might include the signal synthesizers used in the modulation and IF processes.
4. *Processor and control interface:* This is the brains of the operation. It controls all of the sections.
5. *Memory:* Memory is related to the processor, the EPROMs used to allow programming of the mobile station. In a GSM phone, the SIM card might be put in this category, as it stores information, but it can also be considered part of the user interface.
6. *Audio:* this is where the audio is processed, tones are created in analog, and perhaps the vocoder is in a digital mobile station. Newer digital base station systems actually perform the voice encoding/decoding at the BSC (vocoded voice frames are sent via the backhaul back to the BSC, and vice versa).
7. *User interface:* This is the keypad, microphone, speaker, ringer, LCD, and backlighting of the mobile station. Perhaps some of the alarming and interconnection that controls the base station might be considered in this category.
8. *Antenna:* This is, naturally, the antenna system. In mobile stations, this might also include a duplexer, which allows transmission and reception using a single antenna (essentially, a splitter with isolation between the receive and transmit paths). Base stations usually have separate receive and transmit antennas.
9. *Power supply and accessories:* This is the power distribution circuitry as well as the battery or external power supply accessories (such as a car kit).

Thus, the job of the technician assigned to troubleshoot a radio is to isolate the problem to one of these sections. This could be a difficult task if a technician does not understand the significance of the tests they are performing.

Keep in mind that manufacturers, particularly base station manufacturers, reserve the right to stray from the above formula, and there is no standardization as far as how a radio should be made.

Using this general concept of a transceiver, one can divide tests to be performed depending on the general category in which, when performed, the tests will indicate problems. Of course, tests are not foolproof, and rarely is a test an indication of exactly one area of a transceiver (there are usually several areas that could cause a similar indication). Thus, here is a general grouping of tests:

- *Receiver tests:* These would include a variety of sensitivity (i.e., the lowest level a receiver can receive and demodulate a signal with a degree of quality) tests, which might include injecting interfering signals into the receive path, raising the noise floor, or simply finding the absolute sensitivity. These tests are usually a test of the receiver front end, the *intermediate frequency* (IF) sections, and the demodulator.

- *Transmitter tests:* These would include a suite of power tests, including testing power output at various power levels, examining the burst, using spectrum analysis to look for spurious and harmonics, and other similar output power tests. These power tests generally test the power amplification process, although, again, they can also indicate problems in other sections, such as the modulation or control sections.
- *Modulation tests:* This is where the modulator is tested to ensure that the voice and data are being placed on the carrier correctly. They might include peak deviation tests, where we look at the AMPS signal to ensure the FM is not exceeding set limits. In NA-TDMA, the primary test would be EVM, which relates to how accurately the modulator is changing the phase and magnitude as it represents the bits. GSM uses only one component of EVM, phase error. In CDMA, we can test EVM as well as a separate modulation test, RHO. These tests are a test of the modulation process, but, again, they can indicate problems in the transmit path or control section as well.
- *Timing and frequency tests:* In digital systems, timing is a critical test, as without correct timing, synchronization is very difficult. This relates to time-alignment tests in TDMA systems, and similar tests called time offset in CDMA. Frequency is also important in all systems. The basic frequency test involves a frequency counter, which indicates the frequency error.
- *Call processing:* Running a mobile phone or base station through the motions it might go through during a call, such as originating a call, performing a handoff, registration, access, authenticating, and such ensures that the control section, memory, user interface, and the general operation of the mobile station is working correctly.
- *Antenna tests:* The antenna represents the transmission path; if it is not working correctly, of course the system will be down. Testing antennas requires some form of scalar network analysis. Essentially, a signal is sent up the antenna, and a measuring instrument then measures any signal that is not passed out of the antenna but rather is sent back down. This return loss is used to calculate a ratio called *voltage standing wave ratio* (VSWR).
- *Audio tests:* Ensuring that the audio section is working properly is another very important test, as the whole purpose of cellular radio is to provide voice communications. This is usually done by setting up a loopback using a communications test set: a technician can talk into the microphone, the voice is encoded and transmitted, the test set loops the audio back, and the voice is played back to the technician via the mobile station's speaker a few seconds

later. This is to a large degree a subjective test, but objective voice quality measurements are very expensive and complex, making them impractical for field use.

10.4 Measuring Power

Probably the most basic of all measurements is the measurement of power. Power measurements may seem rather simple, but there are several factors that make a power measurement a bit more complex than it seems on the surface. It is important to understand there are several different ways to measure power, and choosing the wrong method can lead to large-scale errors that seriously affect network performance.

In all of the cellular formats, power in a particular channel, generally the control channels, is what limits the size of the cells. Increase the power of the base station and the cell expands, decrease it and the cell shrinks. The size of the cell, of course, is carefully calculated by performance and optimization engineers to ensure both proper coverage and the prevention of cochannel interference (the use of the same frequency by two base stations) or PN pollution in the case of CDMA (too much signal with a different PN offset in a cell, the equivalent of a high noise floor). The mobile station's power is equally controlled by the network for similar reasons. If power levels of both systems (the base stations and the mobile stations) are not calibrated properly, the entire system can begin to have very bad interference, holes in coverage, and generally poor performance (see Figure 10.2).

The peak power meter was used in the early days of radio. A peak power meter finds the peak power of the signal and reports it. This is needed to find the maximum transmit power output of an AM radio station. It can also be used in many applications on systems that have a constant power envelope, such as FM and GSM, as the peak power will normally equal the average power. Using a peak power meter to measure signals that do not have a constant signal envelope can be quite inaccurate. In particular, measuring a CDMA signal with a peak power meter will generally lead to higher than actual readings.

The reason a peak power reading of a CDMA signal will be significantly higher than the average power relates to what is called the *crest factor* (see Figure 10.3). The crest factor refers to the waveform and how it is somewhat noiselike in that its power varies randomly. If you look at the top of the CDMA signal (amusingly called the *bart-head*), you can easily see how it looks like noise. Specifically, the crest factor is calculated by dividing the peak power by the average power. Naturally, if a crest factor is high, as it is in CDMA, a peak power reading will be completely inadequate.

Average power is self defining. It amounts to the average amount of power in a signal over a defined sampling period. As mentioned earlier, this is the method needed to accurately characterize the power output of CDMA and NA-TDMA

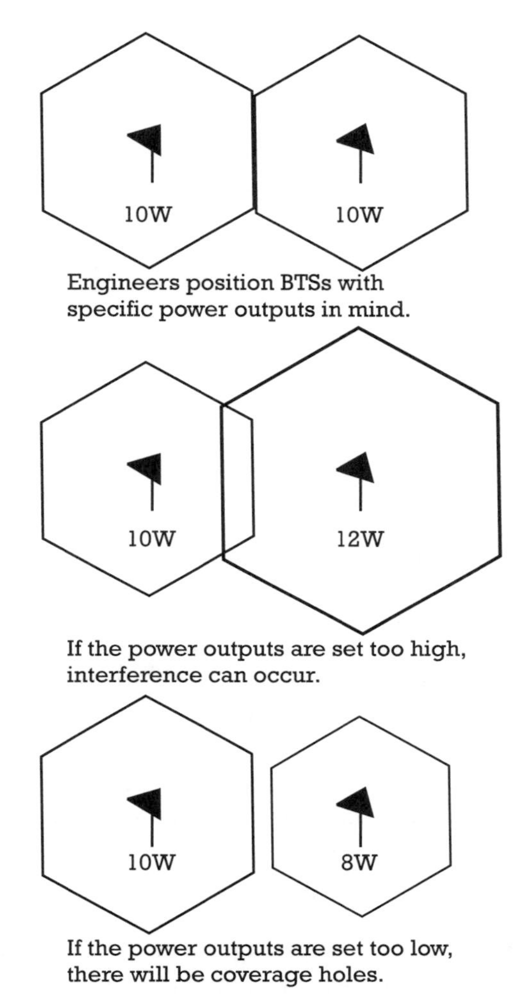
10W
10W
Engineers position BTSs with specific power outputs in mind.
10W
12W
If the power outputs are set too high, interference can occur.
10W
8W
If the power outputs are set too low, there will be coverage holes.

Figure 10.2 Improperly measured and adjusted power can lead to severe performance issues for a network.

systems. One type of power meter uses thermal power sensors (sometimes called *power heads*). These power meters can determine the amount of heat generated by the signal they are measuring and quite accurately convert this heat reading to watts.

Another type of power meter uses a diode sensor, thus converting the signal to a dc value and computing an average. The diode sensor can make measurements at much lower values but will have trouble making measurements at certain levels because the power level is out of the *square-law* region of the sensor. Oversimplifying, this relates to the characteristics of an RF detector and how the sensitivity will drop off whenever the input power exceeds a certain level (typically –20 dBm).

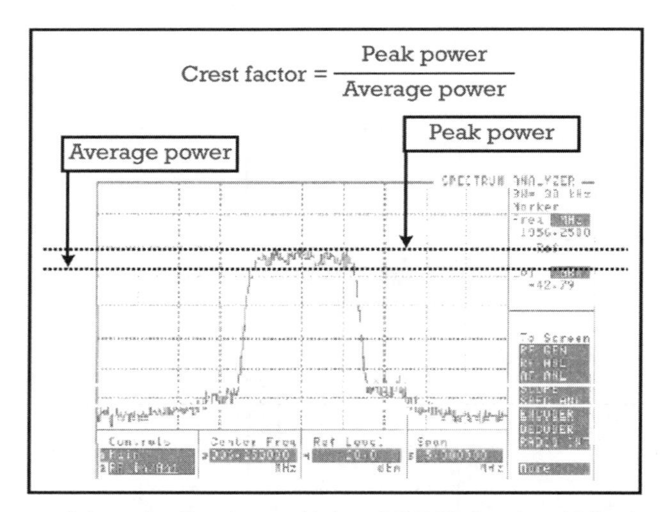

Figure 10.3 Because of the noise-like characteristics of CDMA signals, which causes a high crest factor, it is important to measure average power and not peak power.

An important consideration when using a traditional power meter is that these are broadband instruments. This means that they are measuring all of the power present in the frequency band in which the sensor is designed to operate. This means that if you hook up a power sensor to the output of a base station transmitting multiple carriers, the power meter will return the average total power on all of the carriers and thus will not be frequency selective. Additionally, if there is error power present out of band (e.g., a spurious signal), this too will be measured and added to the measurement and could cause problems.

To compensate for these problems, channel power can be measured. Channel power is a measurement of the average power in a specific bandwidth. Used particularly in CDMA, channel power meters typically down convert the signal to an IF, sample the signal, and perform A/D conversion. The digital samples can then be put through a complex digital signal processor to produce the channel power reading. This type of measurement is critical in CDMA systems where there are multiple CDMA carriers present and a technician needs to measure the power in a particular carrier at a time. It also tends to be more accurate at lower levels than a standard average power meter because the amount of noise introduced into the measurement can be limited by looking at a smaller bandwidth. Additionally, any out-of-band error noise will not be added to the channel power measurement.

Several other measurements relate to power, which will be discussed in greater detail later in this book. These relate to examining the power levels in different domains (remember domains from Chapter 1?). In particular, spectrum analysis is a power measurement in the frequency domain, where the amount of power in specific frequencies is shown graphically. Time-gated power is a measurement made in

TDMA systems. Because the TDMA systems examined in this book (GSM and NA-TDMA) use bursting techniques, it is important to examine the power during the burst to ensure that the ramp up, ramp down, and top of the burst meet a specification. Code-domain power is a CDMA power in which the amount of power in each Walsh code channel can be displayed graphically. There are several other power-related tests as well, which again will be covered in the next four chapters (see Figure 10.4).

10.5 Time and Frequency Error Measurements

While many of the specifics in measuring (in particular timing, but frequency error as well) will be covered in the next three chapters, some of the concepts are generic to all communication formats.

Testing frequency error requires an absolute standard. Frequency standards are maintained by national and international agencies to ensure that frequencies are as standard as time (i.e., 100 MHz is 100 MHz for all systems). CDMA uses extensive frequency and timing standards by using a satellite connection (GPS) at each base station to ensure that all base stations use exactly the same timing and frequency standard. GPS frequency and time standards are becoming inexpensive enough that most repair shops are installing them so that their test instruments will be based on nearly perfect standards. Most test instruments have a 10-MHz standard input/output, so they can synchronize to such an external source.

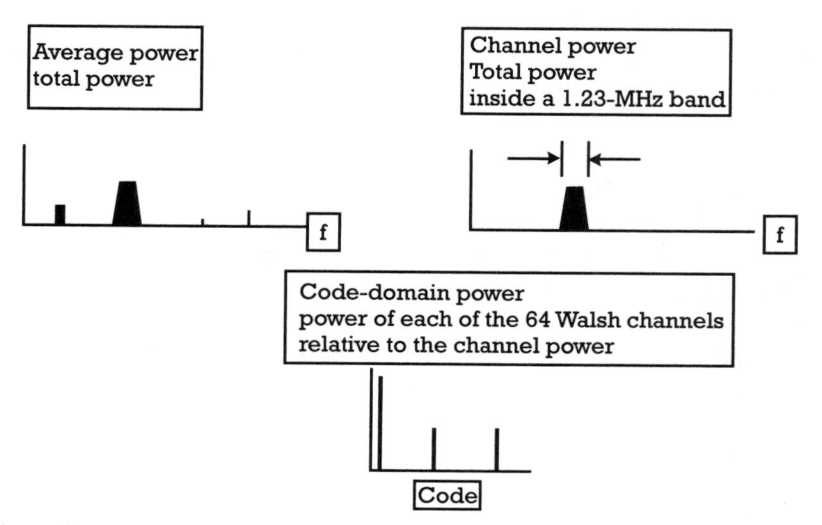

Figure 10.4 Three ways to measure CDMA power, all having their own application. Average power is a broadband measurement, channel power is power only in a specific band, and code-domain power is power within specific codes within a specific band (used in CDMA only).

CDMA uses several timing signals off of the GPS receiver. First, it takes the 10-MHz signal and uses it as a frequency standard. Thus, if you are testing a base station, it is important to take the frequency reference out of the base station and use it in your test instrument to make frequency error measurements. Some CDMA base stations do not output 10 MHz as a reference for testing but rather use a derivative of the chip rate. Most test instruments can take either frequency reference.

CDMA also uses a signal called *one pulse per second* (1PPS). This 1PPS is a synchronized pulse that CDMA uses to synchronize the start of its various messages and channels. A CDMA base station will generally turn the 1PPS into what is called the *even-second clock*, essentially taking every other pulse off the 1PPS. This even-second clock is what is the basis of the time-offset measurement and is thus very important in CDMA measurements.

Thus, problems with timing or frequency may not always be with the transceiver hardware itself but can be with the GPS or frequency standard system the base station is using. Thus, it can be important to include an independent GPS receiver as part of a test package. If the test instrument and the system under test are both using the same frequency and timing reference and the frequency and timing errors pass, yet when separate GPS timing references are used for the same test and the test fails, it is an indication of a possibly bad reference standard.

Mobile stations typically receive timing information from the base station (or test instrument), so a timing reference is not absolutely important in mobile station testing. This is not to say timing measurements such as time alignment tests are not critical, just that they are typically relative to the base station and not based on absolute timing signals, as CDMA base stations are. Frequency references, however, are very important because frequency error in a mobile station can cause problems with reception as well as interference issues, as the frequency channel may begin to overlap into adjacent channels.

10.6 Spectrum Analysis

If you have been an RF technician for any amount of time, undoubtedly you have come across a spectrum analyzer. Spectrum analyzers have for years been the primary tool of RF technicians and engineers—for good reasons. The spectrum analyzer is a very versatile tool that can tell a technician or engineer quite a bit about a system very quickly—if the user understands how to use the instrument correctly. It is very easy to make bad measurements with a spectrum analyzer if you do not understand some of the settings, particularly if the settings are *uncoupled* from each other.

Spectrum analyzers are used to observe signals in the frequency domain (see Figure 10.5). There are several types of spectrum analyzers, but the most common spectrum analyzer is called a *swept tuned* spectrum analyzer. Figure 10.6 depicts a

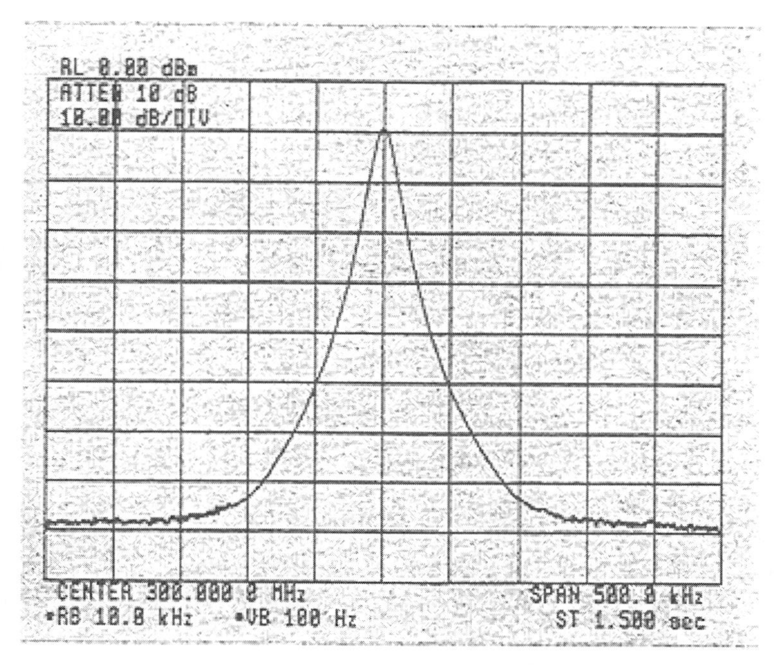

Figure 10.5 A typical spectrum analyzer showing a continuous wave signal.

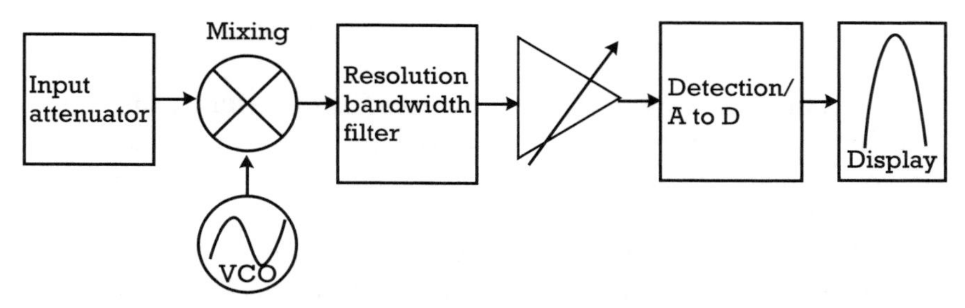

Figure 10.6 The very basic fundamental block diagram of a spectrum analyzer.

typical spectrum analyzer. It is important to understand how a spectrum analyzer works in order to really understand what many of the settings indicate during a measurement.

If you can follow the block diagram, you can see that the input signal is preparing to be measured by going through an attenuator (the front end) and a filter. The signal is then mixed with another frequency from the *voltage-controlled oscillator* (VCO). The frequency of the VCO will sweep through a defined set of frequencies that depend on the span and start/stop or center frequency settings on the spectrum analyzer. The speed of this sweeping can also often be controlled by the sweep time setting.

The output of the mixer is then filtered through the *resolution bandwidth filter* (RBW). It is then put through a detection process that turns the signal into a dc value, which can then be used to plot video. Understanding the RBW function is critical in understanding how a spectrum analyzer measures power. If the signal is wider than the RBW setting, the spectrum analyzer will only measure and show the power based on the level it sees in the RBW filter. This means that if you are looking at a CDMA signal that is 1.2288 MHz wide with an RBW of 30 kHz, the level the spectrum analyzer will show is the amount of power in 30-kHz slices. Because spread spectrum signals like CDMA generally need power measurements in *channel power,* you would need to factor this in to compute a power measurement. This can be computed (in decibels) by the formula:

$$\text{Channel power} = \text{measured decibels} + 10 \log (\text{RBW}/\text{bandwidth}) \quad (10.1)$$

where measured decibels is the marker value at a random point on the signal. RBW is the RBW setting on the spectrum analyzer. Bandwidth is the size of the channel to be measured—in the case of CDMA, 1.2288 MHz.

Thus, if we wanted to find the channel power of a CDMA signal using a 30-kHz RBW, and we obtained a marker measurement of 10 dBm on the top of the CDMA signal, we would compute:

$$\begin{gathered} 10 \log (30 \text{ kHz} / 1.2288 \text{ MHz}) = 16.1 \text{ dB} \\ 10 \text{ dBm} + 16.1 \text{ dB} = 26.1 \text{ dBm} = \text{CDMA channel power} \end{gathered} \quad (10.2)$$

This method is naturally not very exact, as it assumes that the signal will have the same average power as that measured by the marker across its entire bandwidth. If you remember that CDMA has a high crest factor, this could lead to bad measurements.

The RBW and attenuator settings on the spectrum analyzer are very important setup considerations when measuring very-low-level signals. The attenuator at the front end of the spectrum analyzer is designed to reduce the level of the input signals so the input will not overdrive the spectrum analyzer's mixer section. A coupled amplifier after the mixer increases the signal level back to the original level. If this attenuator and coupled amplifier is set up for a large signal (thus, a high attenuator setting), this will dramatically increase the noise level, as the amplifier will magnify any noise present internal to the spectrum analyzer.

Thus, the attenuator should be set as low as possible if the signal to be measured is low as well. The RBW setting is how the spectrum analyzer will show power. If it is set very wide, it will allow more noise power in, and thus the noise level displayed will increase. This makes it harder to make low-level measurements. As the RBW is made more narrowed, less noise is let through, and thus lower level measurements can be made. A third setting, the *video bandwidth* (VBW), pertains to the video displayed. Adjusting this setting has the effect of smoothing out the signal displayed (making noise look less noisy), which can be helpful if the signal to be measured is close to the noise floor.

One of the primary uses of a spectrum analyzer in the field is to find and measure any spurious signals that may be present in the transmitter. Spurious signals are unwanted and generally unrelated in frequency to the carrier. They are measured in dBc, which means decibels referenced to the carrier of interest. To measure a spur, a technician would first find the peak value of the carrier, as any spur found will be measured relative to this signal. If the carrier is a spread-spectrum signal such as CDMA, channel power is generally used as the reference, so again you must take the RBW used into consideration.

Harmonics and intermodulation products are unwanted signals out of a transmitter, but these are related in frequency to the carrier (see Figure 10.7). Harmonics will occur at multiples of the carrier. Thus, if the carrier is 800 MHz, the first harmonic will occur at 1,600 MHz, the second will occur at 2,400 MHz, and so on. Harmonics are measured the same way as spurious, and the measurement is also in dBc. Intermodulation products also occur at certain distances from the carrier frequency. Intermodulation is a process that occurs when two frequencies mix. For the

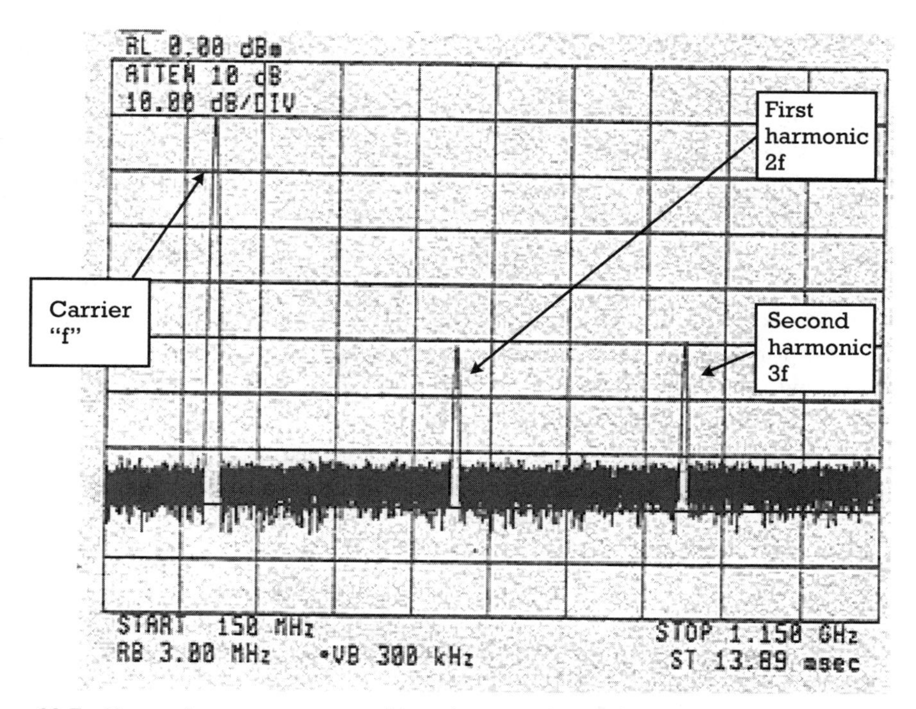

Figure 10.7 Harmonics are common problems in transmitters. They can be found easily by looking at multiples of the carrier.

most part, RF designers do their best to design systems and filters so that these products do not occur at the output of the working system. As systems degrade, however, intermodulation can be a common problem. Intermodulation products are also measured like spurious and harmonics with a measurement in dBc.

10.7 Scalar Network Analysis

One of the more common problems with a network is that the antenna tends to have problems. Because the antenna is subject to all of the elements, failures should not be surprising. Routine maintenance of antennas is very important, as the most expensive base station in the business is still only as good as the antenna system it is using to transmit. The best way to test antenna and cabling networks is to use scalar network analysis. Today, such scalar analyzers are getting smaller and smaller, to the point that several models can fit in small briefcase—a far cry from the early days when the standard network analyzer was a two-man lift.

The simplest application for scalar analyzers is measuring cable loss. Basically, the scalar analyzer generates a swept frequency signal that can be received and measured. To set up for the test, the output of the sweep generator is input into the analyzer, and the frequency response is normalized. When this is done, there should be reference line indicating 0 dB. Now, the frequency response of any device (e.g., a cable) placed between the sweep generator and the analyzer will be shown on the analyzer. Sharp dips in this measurement can indicate problems with cables.

When measuring antennas, the typical measurement is the VSWR. VSWR is a mathematical conversion of return loss, which is easily measured using the scalar analyzer (see Figure 10.8). In this test, the output of the sweep generator is input into an RF bridge or directional coupler. A directional coupler can separate the RF energy being transmitted from the RF energy reflected back. It is this reflected energy that will be fed into the analyzer. Thus, the directional coupler or bridge will have three ports: an RF input, a coupled output that goes to the analyzer, and an output port where the device will be placed for test.

All reflected power will be sent to the analyzer. Return loss is a relative measurement, so it is assumed that 0 dB of return loss means that all of the power is reflected. This can be performed to normalize the analyzer and set up a reference by first placing a short on the output port, which when measured will become the ground reference. Then an open should be placed on the analyzer and the analyzer referenced again. This reference level then indicates 100% of the energy being transmitted. Now, the antenna can be placed on the output port in place of the short. Any difference between the reference and the antenna will be the return loss measurement. This can then be converted into the VSWR measurement using a standard table (see Appendix B).

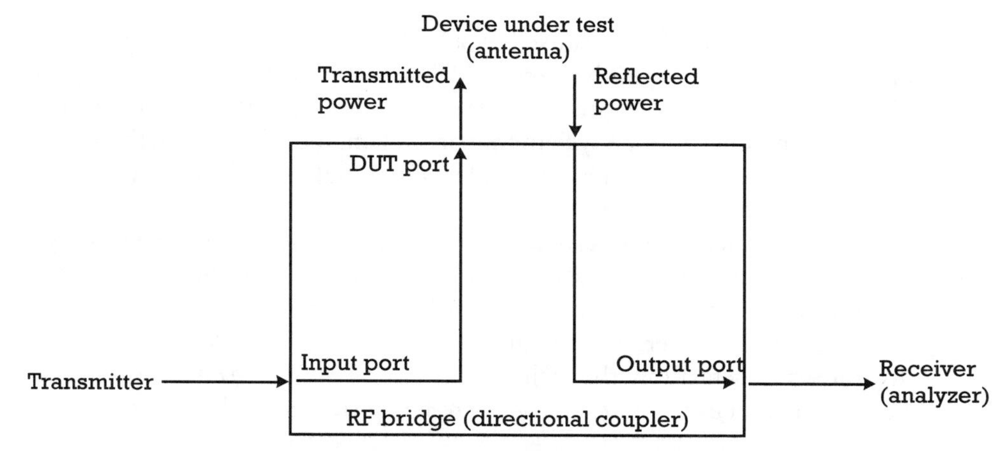
Device under test
(antenna)
Transmitted power
Reflected power
DUT port
Input port
Output port
Transmitter
Receiver
(analyzer)
RF bridge (directional coupler)
Step 1: Calibrate the system by first placing a short and then an open on the DUT port, making this the reference.
Step 2: Connect the antenna to the DUT port. The difference between Step 1 and this measurement is return loss.

Figure 10.8 Scalar analysis makes us of the RF bridge or directional coupler, which allows measurement of reflected power.

It is important to understand what the return loss measurement is telling you. If too much energy is getting reflected, it indicates possible a bad match on the connectors or a bad antenna. Most of the energy should travel out the antenna and not be reflected back. Also, because this measurement is performed by sweeping the frequencies across the frequency used by the base station or mobile, the return loss should not have severe variations in one particular frequency range, as this could indicate a mismatch condition as well.

Many spectrum analyzers have such scalar network analysis systems built in, called *tracking generators*. This is usually very handy for the technician, as they can bring less equipment to the field with them. Of course, an external RF bridge is still needed. As mentioned earlier, newer scalar network analyzer designs are also available in much smaller packages. These network analyzers usually have the RF bridge built in. Many of these newer systems also have distance-to-fault capabilities. This process uses the return loss method, along with some mathematical computations to show technicians precisely where problems with a cable system may be. This makes troubleshooting much easier.

10.8 Testing the Audio and Mechanical Components

Every cellular phone has a speaker, microphone, and user interface, and these items are often the cause of problems in phones. Audio quality is a chief complaint of many customers. It often is caused by a bad microphone or speaker, which is generally an easy and inexpensive repair. While audio analysis gets quite complicated with digital modulation systems, for the most part, even the newest handsets still use a speaker and microphone system that are fed analog voice. Also, the keypad used to send DTMF tones also can have problems that make it difficult for the user to make calls or use special features.

More advanced audio tests will be discussed in the specific technology test chapters. A communications test set can, however, put cellular phones into a test mode in which voice spoken into the microphone is looped back with a delay, which then can be heard in the speaker. This ensures the audio loop is working correctly. Additionally, most test sets allow the technician to press the DTMF keys on the phone, and the test set can ensure that the tones are being transmitted correctly.

10.9 Testing Modulation

Phase can best be represented in the form of a circular diagram known as a *constellation diagram*. In this diagram, the phase of the signal is represented by the angle around the circle, while the amplitude of the signal is represented as the magnitude away from the origin (or center) of the circle, with the very center of the circle representing the least amount of amplitude. As discussed in the previous chapters, all of the main cellular and PCS systems use what is called *constant amplitude phase modulation*, meaning that the amplitude of the signal stays constant during the decision points, changing only (if at all) during phase transitions. Thus, each decision point has two components: phase and amplitude.

Deviations from the ideal vector for each decision point is known as the EVM, which is a combinations of two submeasurements: phase error and magnitude error.

In the three widely used cellular and PCS systems in use today, three different types of phase modulation are used. In GSM, GMSK is used. In IS-136, $\pi/4$ DQPSK is used. In CDMA, QPSK and O-QPSK are used for the forward and reverse link, respectively. Each can use EVM as a good gauge of modulation quality. However, as GSM uses a constant envelope, amplitude error should not be a factor; thus, just the phase component, known as *phase error*, is typically measured.

10.10 Introducing the Communications Test Set

A spectrum analyzer, power meter, frequency counter, scalar network analyzer, and modulation meter, among other instruments, are needed to completely test a radio system, be it the mobile station or base station. Naturally, carting around five to seven different instruments is really impractical. This became particularly evident as radio sites started getting located at very remote sites, such as oil rigs, mountain tops, and skyscrapers. Thus, several companies began combining these instruments into one single instrument, which came to be known as a service monitor or communications test set (see Figure 10.9).

Today such test sets are the common tool of the technician, combining all of the instruments technicians might need in a field-ready package, ranging from 30 lbs to 60 lbs and getting lighter with every new generation of product. Originally, these test sets were simply spectrum analyzers with some added features for testing modulation. New generations of the instrument are much more than this, incorporating high-level testing capabilities for the digital modulation formats.

The typical service monitor includes a spectrum analyzer, power meter, frequency counter, deviation meter, distortion meter, audio oscilloscope, digital modulation testing systems, SINAD meter (along with a reverse channel generator), as well as other typical field service tools, all of which vary from test equipment vendor to vendor.

10.11 Drive Test Tools

A variety of test tools are used to evaluate and optimize cellular networks. Optimizing coverage and capacity is becoming more important as competitive forces increase, particularly in cellular markets in the United States, where as many as seven different cellular and PCS carriers vie for subscribers.

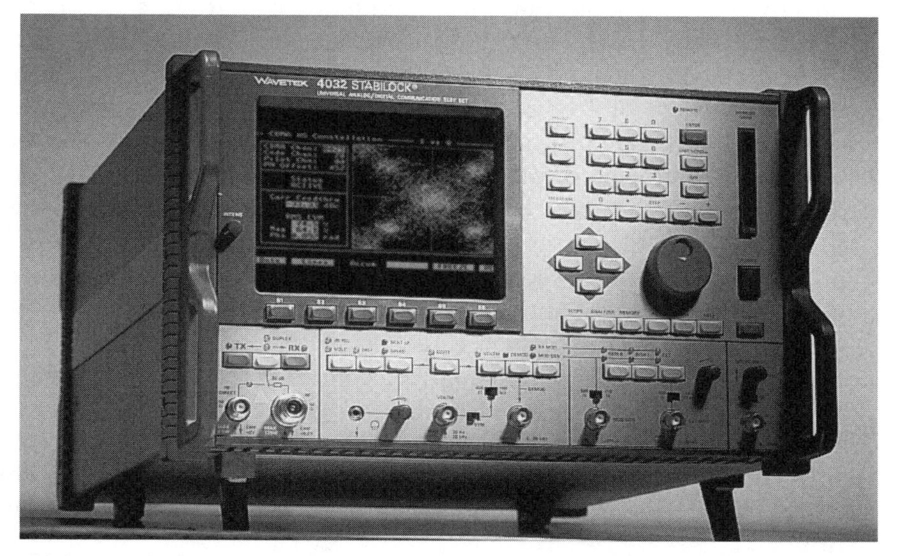

Figure 10.9 A typical communications test set, like this one, incorporates several instruments in a compact package, saving the technician from carrying several instruments to hard to reach sites or using up too much bench space.

Most of the tools used for these purposes are a derivation of the field strength meter—a tool that can record the field strength accurately and usually plot this measurement into a mapping software program. These systems have become quite advanced in recent years. Today, the systems can extract the field strength information from cellular phones and, using sophisticated GPS mapping software, print extremely accurate and concise maps of not only the carrier's coverage, but very often the competitor's coverage as well. Such receiving systems can also be used in tandem with base station signal generators to test for the best antenna orientation or other such propagation issues before the base stations even arrive—or perhaps to decide whether another base station would help coverage at all.

Along these same lines are other instruments, such as voice quality systems that can give an objective measurement of the overall voice quality in a network. Also, as capacity becomes an important consideration, systems that can create heavy loading on a cell and then measure the cell's overall performance are also common.

10.12 The Battery

Any technician who has ever spent time working the service counter has probably heard the complaint from customers that their phone's battery doesn't last as long as it should and thus must be bad. A variety of specialized test equipment has developed to help determine if batteries are truly degrading or if perhaps the phone has a problem with its power supply circuitry or in its RF circuitry. Remember, how much and how high a phone transmits, sleep-mode operation, and other performance parameters can all affect how much current a phone will draw.

As battery quality and performance improves, the battery will probably become much less a cause of problems. For instance, many mobile station users will place their phone or battery in a charger whenever they aren't using it. This can lead to the battery degradation, as the battery was designed to be completely drained before it is placed in a charger. Newer lithium ion batteries generally do not have this problem. However, many older batteries are still in the market, and cost is a factor. Thus, several power system tools, such as battery reconditioners, battery drainers, and go/no go testers, have evolved and have become standard in most phone shops.

Poor battery life can often be caused by network conditions. For instance, if digital service is poor, and the user's mobile continually drops to AMPS mode, the battery will very quickly lose its charge. Additionally, handset power draw (and therefore battery life) is directly related to handset transmit power. Thus, if a handset is in poor coverage areas, or a base station has a poor receiver condition, the handsets will have to transmit at higher levels and will use more battery power than normal.

10.13 Conclusion

There is no way this text can be exhaustive regarding test methods on the various formats. However, with a basic understanding of the technology, a technician or engineer should be able to quickly understand what a method they encounter is designed to test.

CHAPTER 11

Testing AMPS

11.1 Introduction to the AMPS Transceiver

Figure 11.1 shows a typical simplified FM radio transceiver. For the most part, this layout has remained similar since the beginning of cellular radio, although each manufacturer will have its own methods. With the advent of dual-mode phones (phones that operate in a digital format in addition to AMPS), this block diagram may look different, but for the most part, all of the components shown will be part of the system. This diagram is the typical AMPS mobile station; however, for gaining an understanding of testing techniques, this diagram is similar enough to the base station.

A base station will consist of multiple transceivers and will contain links to a central controller and the switch. Also, many base stations will use common components for multiple transceivers, such as the antennas, duplexers, amplifiers, and filters.

The transceiver consists of a receiver path and a transmit path. Both share a common microprocessor, interface (such as the keypad on the mobile station or the connection to the switch in a base station), and frequency synthesizer section. The transmit and receive paths are for the most part an inverse of each other. The audio sections of the receiver and transmitter were discussed in detail in Chapter 3. Naturally, any limiting, companding, and preemphasizing that is done in transmission should be done on reception to ensure that the effect on the transmitted voice is as little as possible.

11.2 The AMPS Receiver

The AMPS receiver is not unlike any common FM receiver—essentially, it consists of filtering to prepare the received signal to be demodulated; a mixer that will output an IF, which is again filtered; a detector section that will output audio or detect SAT or wideband data; followed by the audio section and speaker.

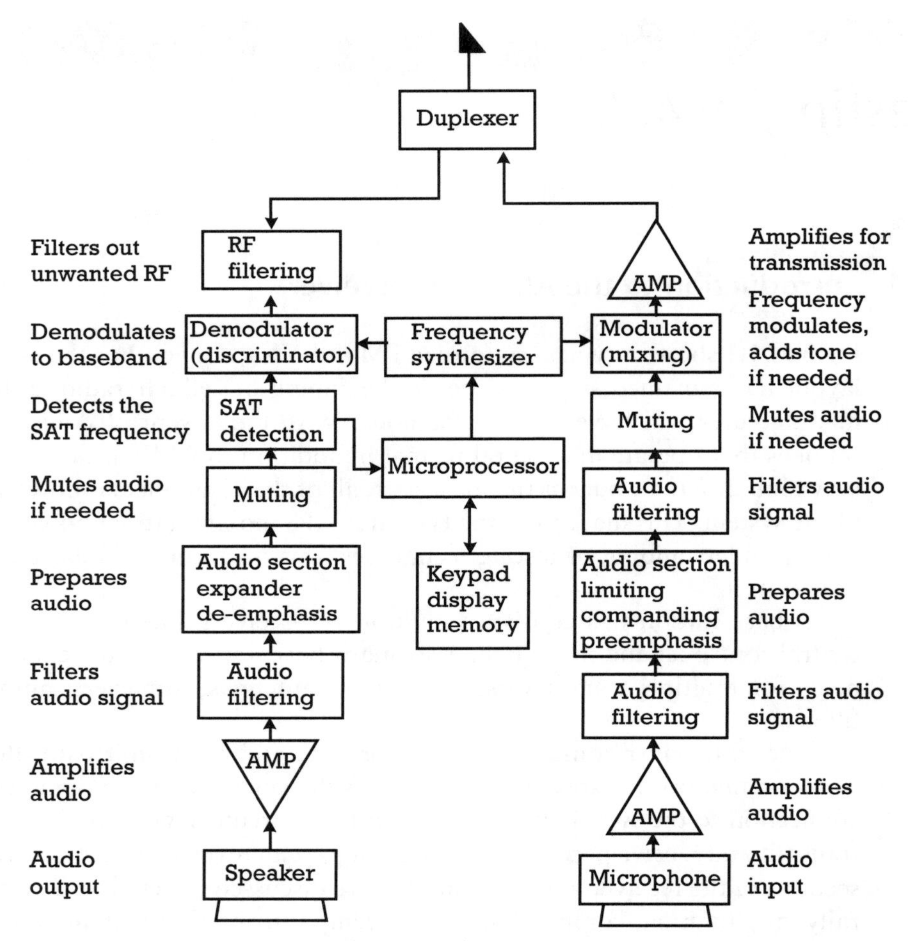

Figure 11.1 The basic components and functions of an AMPS transceiver.

As mentioned earlier, components that may be shared with the transmitter include the antenna (for mobile stations), the duplexer, the frequency synthesizer, and the microprocessor (for the rest of this text, this includes the memory section).

Six analog receiver tests are commonly performed:

- *Sensitivity:* This is a test of the lowest level the receiver can "hear." Usually, it is determined by the front end of the receiver (the initial amplification of the received signal as well as initial filtering) in addition to the IF section.
- *Baseband distortion:* This is a distortion of the audio usually caused by signals harmonically related to the fundamental frequency. They are caused in the receiver system (usually either the frequency conversion or baseband

processing sections) by leaking signals mixing due to nonlinearity in the receiver circuitry (multiple signals input into nonlinear devices will mix).

- *Hum and noise:* This is very similar to baseband harmonic distortion, except it pertains to distortion not related to the fundamental frequency. Hum relates to the low frequency noise that is usually coming from the power supply of the phone.
- *Audio response:* This is a test of how the various audio frequencies can be recovered. The frequencies that cover the human voice must have a good frequency response. Audio response is a test of the overall audio section.
- *Selectivity:* This is a measure of how well a receiver can demodulate the wanted signal while rejecting an adjacent signal. It is a test of the receiver's filters, which should reject these adjacent signals.
- *Intermodulation immunity:* This is a measure of how well the receiver can guard against a degradation in the receiver output in the presence of two or more unwanted signals (which will mix in the nonlinear environment of the receiver). It is more of a production test; however, as it does simulate the real world, it is performed in high-level service centers. Generally, it is a test of the nonlinearity of the receiver's front-end circuitry.

11.3 Filters and Testing AMPS

Many of the receiver tests are actually tests of the audio signals output by the receiver. An important consideration is what is called a *psophometric filter*. These are filters meant to match the frequency response of the output signal to that which the human ear can hear (see Figure 11.2). The human ear can perceive differences between audio signals that may have a lot of high-frequency noise and audio signals that do not. For this reason, when testing, a test engineer may want to eliminate such high-frequency noise from the test. This would be performed by placing such a filter inside the test instrument. In European analog systems, a filter often called CCITT was prescribed for all such testing. The Americas never specified such a filter, but sometimes they are implemented on a carrier-to-carrier basis.

11.4 Testing the AMPS Receiver

Tests for the AMPS receiver include:

- *Sensitivity:* Testing sensitivity involves determining the lowest RF level a receiver can demodulate and still produce a usable output. In analog, this

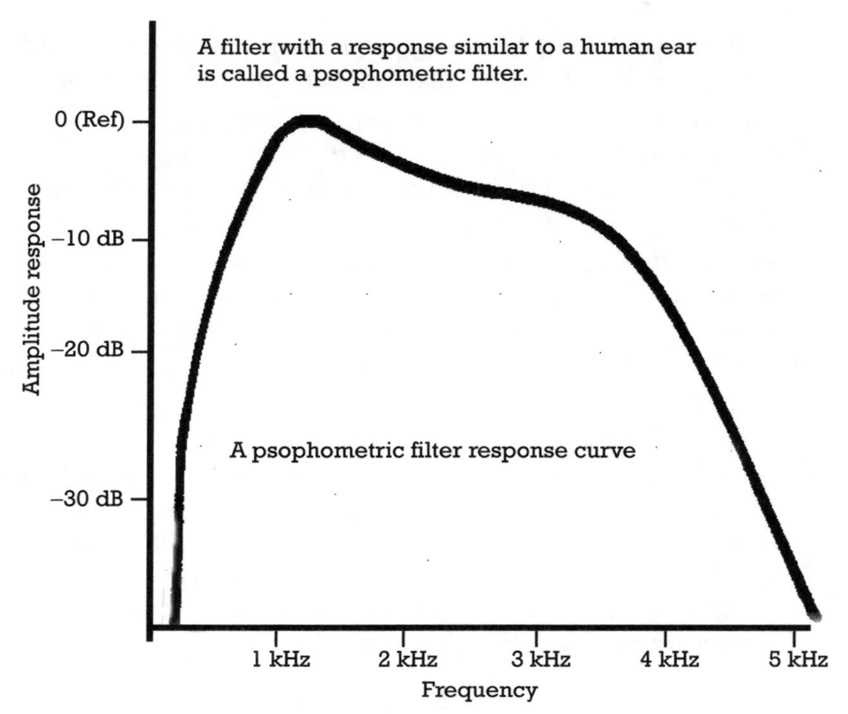

Figure 11.2 A psophometric filter is often used in the test instrument to make sure all of the tests do not take into account interference from audio frequencies that have no effect on the human ear.

usable output is determined using the SINAD measurement (see Figure 11.3). SINAD is a ratio (measured in decibels) of:

$$\text{Signal} + \text{Noise} + \text{Distortion} / \text{Signal} + \text{Distortion} \quad (11.1)$$

The signal, noise, and distortion of the receiver generally remain constant, regardless of the input signal (as the signal level goes down, the noise level will increase). Thus, as the RF level is decreased, the audio signal and distortion will reduce. When the audio signal and distortion are 12 dB lower than the signal, distortion, and noise level, then the RF level being input into the circuit is recorded. This is the 12-dB SINAD measurement. In most systems, this specification is –116 dBm (see Figure 11.3).

Measuring SINAD is a bit complex, but most communication test sets have the function built in. The setup requires an RF signal with a tone modulated on it (1 kHz) to be input into the receiver. As the SINAD is a measurement of the audio, it requires the audio to be separated and input into the test instrument. This is usually performed by a *break-out box*, which essentially

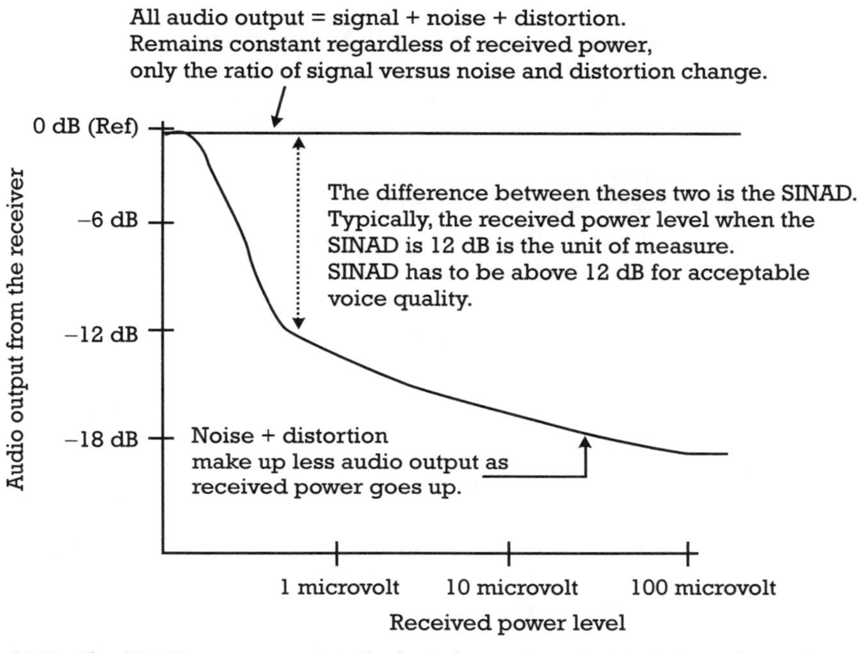
All audio output = signal + noise + distortion.
Remains constant regardless of received power,
only the ratio of signal versus noise and distortion change.
0 dB (Ref)
-6 dB
-12 dB
-18 dB
Audio output from the receiver
The difference between theses two is the SINAD.
Typically, the received power level when the
SINAD is 12 dB is the unit of measure.
SINAD has to be above 12 dB for acceptable
voice quality.
Noise + distortion
make up less audio output as
received power goes up.
1 microvolt
10 microvolt
100 microvolt
Received power level

Figure 11.3 The SINAD measurement is the basis for most receiver tests in analog radio.

interfaces with the transceiver and allows for the audio to be output separately as the transceiver operates on a call with the test instrument.

The measurement is thus on the audio output of the receiver. A SINAD meter is essentially a voltmeter with a couple of twists. First, the voltmeter (actually a distortion meter, but it is essentially a voltmeter for voice frequencies) is used to measure everything being output from the receiver. This is the signal, noise, and distortion. Then, because we know the signal's frequency (1 kHz), a notch filter is used to get rid of the energy at this frequency. This is the noise and distortion. Subtract the two, and you have the SINAD measurement (remember, dividing logarithmic functions such as decibels is performed by subtraction).

Again, thankfully, most communication test sets perform this measurement automatically. All that is required is the RF signal, modulated with the 1-kHz tone, to be input into the receiver, and then the audio to be broken out of the receiver and input into the test instrument.

- *Hum and noise and receiver baseband distortion:* These are tests of how the audio may be getting distorted with unwanted audible noise. The test is actually very similar to SINAD. Generally, a test instrument called a distortion analyzer is used, with the same setup described for SINAD. The 1-kHz signal is modulated and input into the receiver, and then it is notched out in the test instrument. The noise, whether it is a low frequency hum or harmonically related, can then be measured.

- *Receiver audio frequency response:* This is a test of how well the receiver can demodulate various audio frequencies. Generally, a reference value is established (at 1 kHz), where the audio output level is measured. The test instrument will then modulate various frequencies, and similar measurements will be made. The response should be flat, with a roll off at frequencies above and below the desired voice frequencies (typically below 300 Hz and above 3,200 Hz).
- *Selectivity and intermodulation immunity:* These are fairly high-level service tests, primarily because they generally need a high-quality signal generator in addition to the communications test set. This makes these tests somewhat expensive (and cumbersome) to be performed in the field. These tests are similar to the sensitivity test, but instead of lowering the RF level of the carrier, the interfering signals are increased in power until there is an increase in distortion. The receiver should be able to tolerate a certain level of interference from these signals (see Figure 11.4). In the case of adjacent channel selectivity, the interfering signal is placed, quite naturally, in the adjacent channel. In the case of the intermodulation immunity, the interfering signals are placed at specific offsets from the carrier to cause an intermodulation product in the channel of interest.

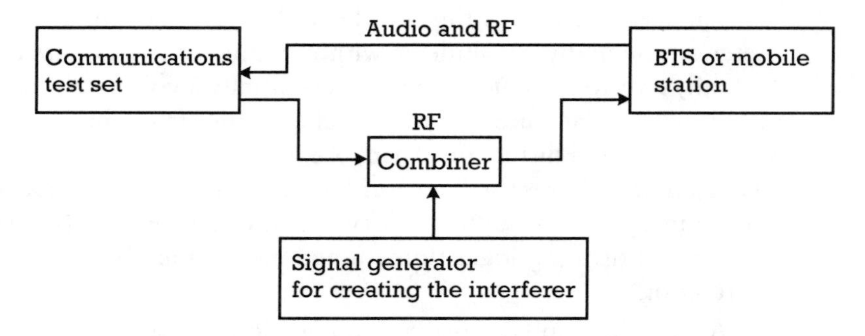

1. The test set generates the transmit signal.
2. The transmit signal is combined (not mixed) with an interfering signal.
3. The mobile station or BTS receives the signals.
4. The test set performs the SINAD measurement via returned demodulated audio.

Figure 11.4 Selectivity, the ability of the receiver to select the correct frequency and demodulate in the presence of interferers, generally requires one or more signal generators to generate the interfering signals.

11.5 The AMPS Transmitter

The AMPS transmitter is also very similar to any FM radio transmitter. Perhaps the main difference is that the audio is processed a bit more, and circuitry has been added for the signaling tones. The primary transmitter tests in the field can thus be divided into the following groups:

- *RF section tests:* These are tests that will test the basic RF parameters. These include frequency error, output power, FM deviation (whether the FM is limited), and spurious.
- *Audio section tests:* These include the proper operation of the microphone, speaker, and audio response (whether all of these audio frequencies are being transmitted correctly).
- *Digital signaling:* These tests are primarily performed on mobile stations. The tests include performing many of the call-processing tasks, such as handoffs, registrations, and authentication testing.
- *Tone testing:* These are the tests on the two tones that are transmitted in AMPS, the SAT and the ST. The tests ensure that these tones are transmitted correctly—with the correct frequency, deviation, and time.

11.6 Testing the AMPS Transmitter

Tests for the AMPS transmitter include:

- *Power and frequency:* Chapter 7 discussed frequency and power measurements in detail. In analog cellular, a basic average or peak power measurement can be made using a typical communications test set. In base station testing, this is typically performed at maximum power and adjusted accordingly. On mobile stations, the power should be checked at each of the power levels the mobile station supports. Typically, problems with power tend to be caused by the power amplifier at the end of the transmit path, although degradations anywhere in the transmission line can cause such a problem. The duplexer can also be a cause of power problems, particularly if the power problem occurs at specific frequency ranges (indicating some sort of failure or degradation of a nonlinear component). Frequency accuracy is controlled by the frequency synthesizer section and could be the result of a bad mixer, a bad local oscillator used for mixing, or perhaps the control circuitry.
- *Deviation:* Ensuring that the FM signal does not exceed the bandwidth allowed (remember, the channels are 30 kHz) is very important. In AMPS, ±12 kHz is the specification for maximum deviation. This test is performed

using a deviation meter, which is a standard feature of most communication test sets. Typically, a call is established, and the technician can speak a specific message into the microphone so that it is transmitted to the test instrument (although in a lab environment, a fixed audio level would be transmitted). The peak deviation can then be measured.

- *Spurious:* This is a spectrum analyzer test in which the technician ensures there are no unwanted signals above a specific threshold. These unwanted spurious usually develop in the mixing process, although they can develop throughout the transmission process. The test is performed by looking at the carrier with a spectrum analyzer, using the top of the signal as a reference (this is typically performed on a base station with no modulation), and then ensuring that no signals rise above a specific threshold, relative to this carrier. Spurious can be as important as power and frequency when it comes to the FCC, as unwanted spurious being transmitted outside a carrier's permitted frequency band can result in interference (and usually a fine if the FCC catches it).
- *Speaker and microphone tests:* The quality of the audio signal is quite important, as it is this voice quality that the customers will equate with their overall impression of the service. The microphone and speaker of a mobile station need to be checked for proper operation. This can be performed easily by using a communication test set that can loop back the voice. Thus, a technician can speak into the microphone and then a second or two later hear his voice, ensuring the audio loop is working properly.
- *Transmit audio frequency response:* This test can be tedious to perform, and, for the most part, automated sequences take care of it. The purpose of the test is to ensure the transmitter can correctly modulate and transmit the audio frequencies evenly. Usually, an FM modulation analyzer is used. A modulation analyzer reports the amount of deviation in a percentage. To test the flatness of the audio frequencies, a 1-kHz signal is transmitted and adjusted until the modulation analyzer reports 20% system deviation. At this point, the modulation analyzer can be set in a delta mode, showing the difference from the reference, and the audio frequency can be varied across the specified range.
- *Call-processing tests:* These tests are typically performed only on the mobile station using a communications test set. To perform the various call-processing tasks, the test set must simulate a base station, and it must send and acknowledge messages from the mobile station.
- *Tone testing:* Once you understand the fundamentals about the SAT and the ST, it is easy to see that testing the operation of these tones is essential to ensuring proper operation of the network. This applies equally to base station testing and mobile station testing.

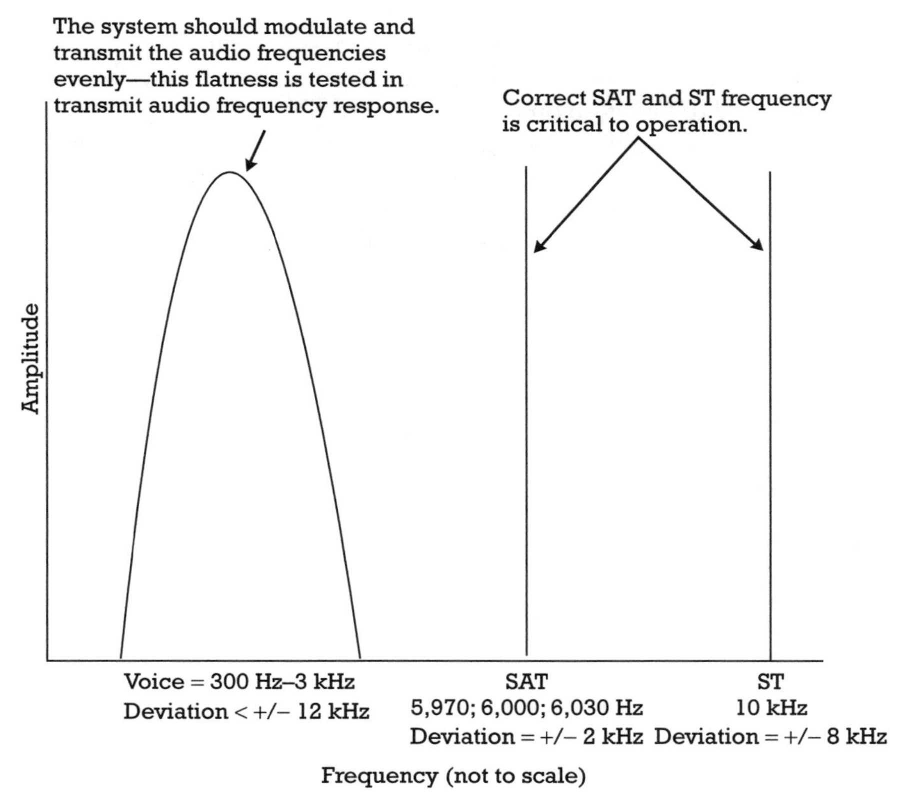

Figure 11.5 The three baseband signals—voice, SAT, and ST—need to have their modulation tested by checking the frequency, amount of deviation, and, in the case of audio, the flatness across the audio frequencies.

Various actions must be initiated by the test instrument (simulating a base station), after which the ST and SAT can be monitored. For instance, a call is originated by the handset, and as the handset moves to the FVC, the test instrument can generate a known good SAT, which the handset should detect, filter, and modulate back on the RVC. This SAT can then be tested for the correct frequency as well as the correct deviation. Generally, the test should be performed three times to ensure that the handset can identify and retransmit all of the SAT frequencies correctly.

Testing the ST requires the test instrument to command the mobile to perform certain tasks or to have the technician initiate an action on the handset. For instance, the instrument might prompt a technician during an autotest to press the send key for a hookflash or press the end key to perform a mobile release; the instrument can then measure the ST and ensure it is at the correct frequency (10 kHz) with the correct deviation (±8 kHz), transmitted for the correct amount of time (in the case of a hookflash, 400 ms; in the case of a mobile release, 1,800 ms).

Testing the operation of the tones on base stations is also very important. If a SAT is not being modulated correctly on a radio, that channel will drop any call that is attempted to be set up on it. The actual testing on the base station is not very different from the mobile. Most base stations can be configured to generate a specific SAT on a particular voice channel. A typical service monitor can then be used to measure the deviation of the carrier (which should be ±2 kHz) as well as the frequency of the tone. This test is performed on each channel and is as essential as carrier power and frequency. Figure 11.5 shows the three AMPS signals that need to be tested: voice, SAT, and ST.

11.7 Conclusion

If you have had experience testing any FM radio, the AMPS cellular phone should be quite familiar. These testing techniques are very similar for all radio systems, including the next generation digital systems. While the methods can change substantially from radio system to radio system, the general concepts remain the same, primarily because the general block diagrams remain very similar.

CHAPTER 12

Testing TDMA Systems—GSM, NA-TDMA, GPRS, and EDGE

12.1 Introduction

This text is taking the approach that TDMA-based systems, whether GSM, NA-TDMA, GPRS, or EDGE, are very similar in their testing needs with regard to RF performance. This is not all to say that these formats do not have significant differences, as the previous chapters have shown. Many of the tests for these digital formats are conceptually the same as tests performed for AMPS systems. For instance, measuring output power may require a different procedure in the test instrument, but the resulting measurement has the same connotation regardless of whether it is a digital or analog system or which digital format it is.

Another important concept to remember is that the NA-TDMA standard (IS-136) is a dual-mode standard. Most NA-TDMA phones (all current production models) are dual mode, meaning they operate in analog or digital mode. Thus, testing needs for these phones include those discussed in this chapter in addition to those discussed in Chapter 8. Dual-mode and multiple-band GSM phones are also beginning to appear on the market, particularly in North America, where such multiple bands and formats are needed to allow international roaming capabilities. Each format and frequency band should be tested as an individual system.

NA-TDMA cellular carriers also utilize dual-mode base stations in order to allow digital and analog service to their customers. The same approach is taken with these base stations—testing digital and analog radios as separate systems.

12.2 Introduction to the TDMA Transceiver

Again, because this text is designed to be a general reference, it will assume GSM and NA-TDMA systems are very similar for testing purposes.

Figure 12.1 shows a typical block diagram for a TDMA system. The TDMA system's block diagram is not terribly different from that of an analog radio; the

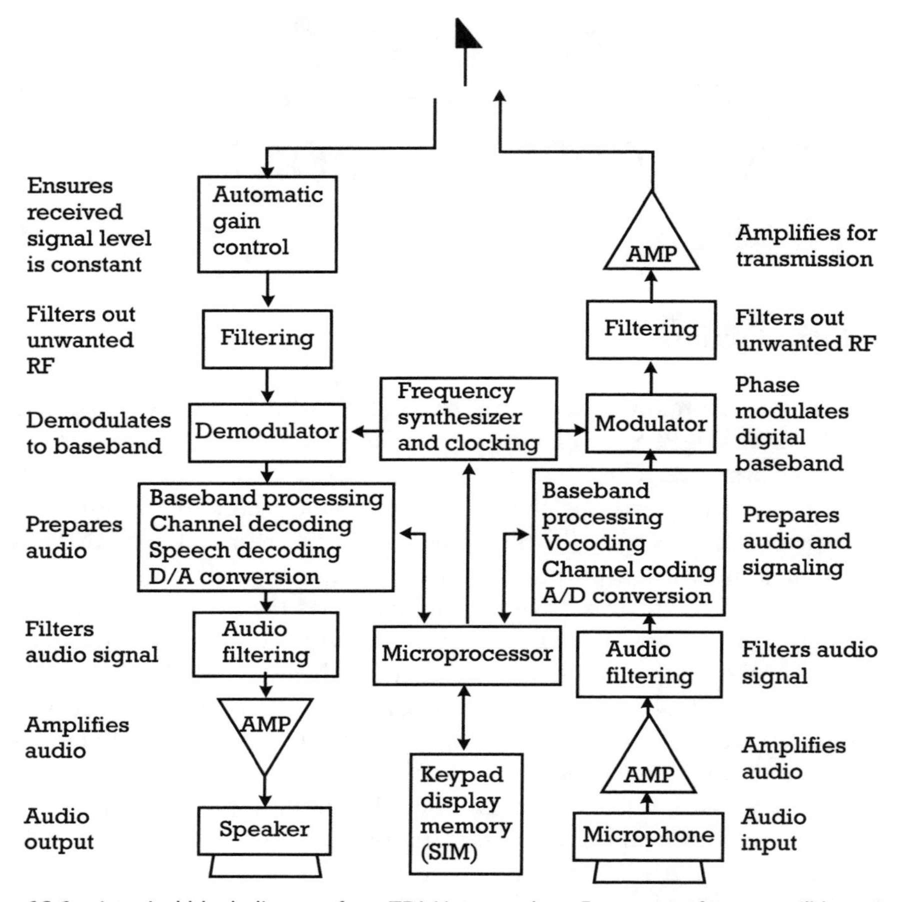

Figure 12.1 A typical block diagram for a TDMA transceiver. Every manufacturer will have its own block diagram, but it will usually be similar to this.

difference, of course, is how the functions are performed. In a TDMA system, these sections include:

- *User interface:* This will be essentially the same for all mobile phones—except that in GSM there is the SIM card interface, which can be used to store user information and interface with the control section for security purposes.
- *Power supply:* A battery is a battery. As most digital phones are more recent designs, they often use lithium ion batteries, which last longer and do not degrade as easily as nickel hydride.
- *Audio or baseband section:* In digital, the analog voice is changed to digital. The audio section includes D/A and A/D converters, the vocoder, and any error correction and detecting coding and decoding subsystems Because much

more information than audio is coded or decoded, this section is often called the baseband section rather than audio.

- *Modulator/demodulator:* In TDMA, these subsystems will take the digital information from the baseband section and modulate it or demodulate it onto or off of a RF carrier. We would also include the local oscillators and mixing sections, which are used to either upconvert the modulated signal to the transmit frequencies or downconvert the received signal to a more usable IF.
- *Transmit section:* The transmit section will consist of filtering to ensure the signal will have the correct occupied bandwidth and power amplification so that it will transmit at the correct level.
- *Receive section:* The receive section consists of an automatic gain control system that ensures that low-level received signals are amplified and high-level received signals are attenuated. This ensures that the demodulator will receive a signal at a continuous level. Filtering will also take place to eliminate unwanted power in the transceiver, which can cause intermodulation products disturbing the demodulation process.
- *Antenna/duplexer:* The antenna and duplexer sections will be similar to those used in AMPS, particularly if the phone is dual mode. If the phone is digital only, it does not require a duplexer because the phone receives and transmits at different times. Degrading performance of duplexers are a major cause of problems in some handsets (generally, their frequency response tends to degrade), so being able to eliminate both the cost and the problems associated with duplexers can be a major advantage for digital phones. An antenna is an antenna, and aside from the fact that digital phone designs are generally newer than analog designs, and thus make use of more advanced antenna technology, the antenna is much the same across technologies. One major difference, however, is that many digital systems are designed to operate on multiple bands, and thus the antenna must be tuned to operate well at different frequencies—a trickier task than one might imagine, as an antenna's frequency characteristics are often the result of length or size.
- *Processor, control, and memory:* Digital systems typically require much more processor and memory resources than analog systems. These sections control much of the complex baseband section, as well as incorporating user interface functions. Several parameters may also be adjustable via the control and processor sections, such as power output, receiver gain, frequency, and IQ modulator balance (the modulation quality). This is usually done via the keypad or a computer connection. These sections are primarily responsible for the "PCS" services that customers demand, such as short messaging, caller ID operation, and the like. While, obviously, the baseband and RF sections must be working correctly as well, the processor, control, and memory sections are what interprets and implements these services. In GSM, the advanced feature

of frequency hopping is also generally a software implementation, and thus it is primarily an operation of this section.

12.3 Introduction to the TDMA Tests

The tests that are required to ensure overall operation generally fall into four categories. Again, the test requirements are very similar to those performed in analog systems, but the method changes significantly. The four categories include:

- *Functional tests:* These are tests of the overall system operation. They typically do not necessarily point to a problem with any section of a transceiver, but relate to the overall system. They also ensure the system has been configured correctly. For instance, if a phone's preferred network was changed by the end user, the user might not be able to get access. These tests include call processing, special services testing, and configuration testing.
- *Transmitter tests:* These are tests of everything from the modulator to the antenna. In a digital system, this includes testing the modulation accuracy. NA-TDMA uses EVM, while GSM uses phase error. It also includes power accuracy, burst timing and accuracy, timing advance, frequency error and spectrum analyzer tests such as spurious, occupied bandwidth, and adjacent channel power.
- *Receiver tests:* Testing the receiver is substantially different in digital systems than in analog systems. This is because the voice data is put through the vocoding process, and thus the traditional SINAD measurement will not work. Instead, it is possible to transmit a known digital sequence to the receiver under test and then measure how well the receiver can demodulate this digital sequence. Any errors are called bit errors, and the amount of bit errors compared to good bits is referred to as the BER. Common field measurements on digital systems include sensitivity, susceptibility, reception in multipath fading, and Doppler shifting and measurement reporting.
- *Audio parameters:* These are the basic tests of the microphone, speaker, and vocoder operation.

12.4 Functional Tests

For the most part, these functional tests are performed in a maintenance capacity only on mobile stations. While network test tools exist which put an overall network through the paces as far as call processing, services and handoff functions, this type of testing is generally reserved for performance engineering, not the typical base station technician.

The following tests are call-processing tests:

- *Origination:* In this test, the technician dials a number and presses SEND on the mobile station. The test set then acts like a base station and should receive the call origination message and process through until a traffic channel state is established.
- *Page:* The tester simulates an incoming call for the mobile station. The mobile station should receive the page message and continue through the process until the alert is sent, causing the ringer to sound. Once the SEND button is pressed, the call is set up.
- *Registration:* The test set sends a registration message to the mobile station, which then returns the information required. At this point, programming can be verified.
- *Authentication:* The authentication process, while performed in GSM and NA-TDMA, is typically tested more often in NA-TDMA. The test set sends challenges and shared secret data update messages to the mobile station, which much respond correctly. Because the A key must be known by the test set, this test is often performed with the correct result being a failure—as the A key should not match and thus the authentication result should not be correct.
- *Short message service:* Basically, this is a test of the operation of the service. Typically, the test set will send a message to the phone, which will then be displayed.
- *DTMF:* Do all of the keys on the mobile station work correctly? In particularly, does the SEND key (called the *hookflash*) work correctly, as this key is used for call waiting and other special services?
- *Caller ID and voice mail indicator:* This test set simply sends the messages for these functions. The phone's voice mail indicator should come on, along with the number of voice mail messages. For caller ID, the number should flash on the mobile station after the message is sent. Typically, a test set allows the technician to set these parameters (e.g., a number for caller ID and the number of voice mail messages).
- *Handoff:* The function of the handoff is very important, as it is during handoffs that most calls are dropped. The test set usually can put a mobile station through a handoff process, making it switch from one frequency to another during a conversation. In NA-TDMA, this also includes handoffs from digital to analog and from PCS frequencies to cellular frequencies.

12.5 Transmitter Tests

The following are transmitter tests:

- *Average power:* Power is, of course, a critical test of the base station as well as the mobile station. If not enough power is transmitted, the range will be limited. If too much power is transmitted, this could cause an interference problem, and it also means the system is working harder than it needs to, which of course uses up battery power more rapidly. Mobile stations should be tested at each power level at which they are capable of transmitting. Base stations are typically measured at maximum power (and then attenuated according to system settings), although GSM base stations use dynamic power control, and often a test program will step it through various power levels as well. Multiple frequencies should be used for power tests as well, to ensure the system's power is the same across its transmit bandwidth.

  Problems with power typically point to the power amplifier circuitry or the calibration tables in the system. Most phone manufacturers have automated systems using communication test sets that can reset these calibration tables. Faulty power output can also often be an indication of a power supply problem, such as a battery degrading.
- *Burst power:* This measurement, better described as transmitted RF carrier power versus time, is usually a time-domain analysis of the carrier power envelope. Remember, in TDMA systems, the system will turn the transmitter on or off in order to fit in the time slot to which it is assigned. If the transmitter is turned on too slowly, data early in the burst might get lost. If the transmitter does not turn off quickly enough, it will interfere with the next burst. Thus, test sets generally use a mask into which the measured burst must fit in order to pass (see Figure 12.2). Generally, if a system fails, this test it is an indication of a bad amplifier or control circuitry.
- *Timing advance:* Mobile stations are commanded by the base station to advance in bits in order to accommodate timing delays due to be different distances from the base station (see Chapters 4 and 5). A test set can simulate this timing advance mechanism and ensure the mobile adjusts accordingly. Problems with timing advance are typically due to control circuitry or the power amplifier section (it is the power amplifier section where power is burst on and off).
- *Frequency error:* Frequency error is a direct result of the performance of the frequency synthesizer transceiver. This test is particularly important in mobile stations. Remember, the mobile station uses the same synthesizer circuitry for transmitting and receiving, and thus must jump quickly between the two

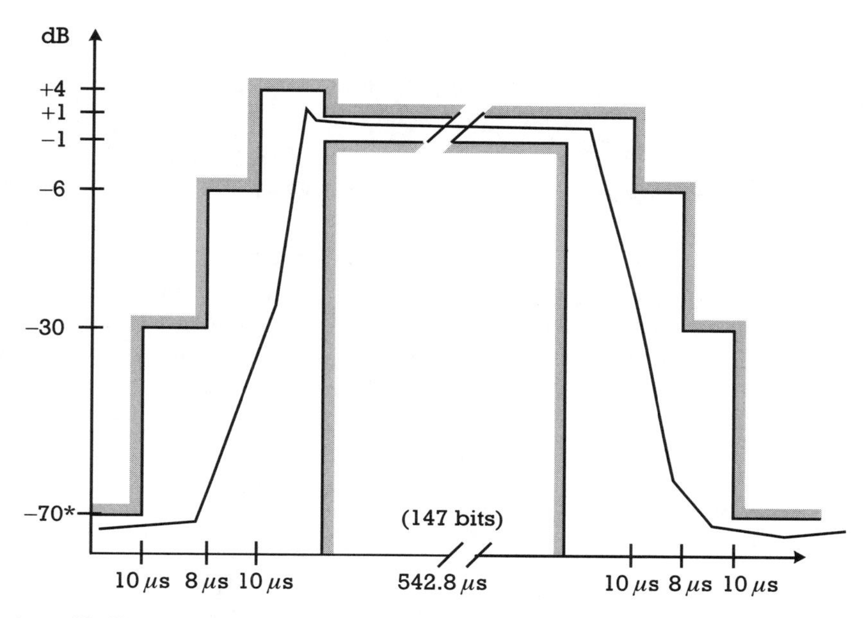

Figure 12.2 A TDMA burst is measured and displayed in a mask such as this. Note, the burst must ramp up and down quickly enough, as well as stay flat across the top, in order to stay within the mask boundaries.

frequencies. Also, in GSM where frequency hopping is employed, the synthesizer must also be able to shift frequencies quickly.

Frequency error will cause other problems, which will show up in other tests. In particular, if the frequency is incorrect, the channel will interfere with adjacent channels. Also, it will make it very difficult for a receiver to lock onto the signal, and thus indications of bad modulation will certainly appear.

- *Phase and amplitude error:* As discussed in Chapter 7, modulation in digital systems can be measured in terms of EVM, which is a combination of the phase and amplitude error. In NA-TDMA, EVM is the primary measurement. This is because NA-TDMA does not use a constant amplitude signal envelope; thus, we must be concerned that the amplitude as well as the phase transitions are correct. In GSM, which uses a constant signal envelope, amplitude error is not as important as phase error. Phase error can be caused by problems with the modulator section, as well as the filters and amplifiers in the transmit path. Amplitude error can also be caused by the same components, although amplitude errors can often be caused by intermittent transmit path connections.

  Graphical representations of modulation often tell much more than just the numeric values. In NA-TDMA, the primary graphical tool will be the constellation diagram. This diagram, which has been discussed in previous chapters, can show patterns that indicate degradations in the modulation and transmission process. For instance, the constellation diagram may indicate a spreading in phase, which might indicate a problem with the modulator (see Figure 12.3). On the other hand, a smearing away from the center might indicate an intermittent solder joint in the transmit path.

In GSM, where phase error is the primary measurement, a graph showing phase versus the transmitted bits is used (see Figure 12.4). This diagram can quickly show deviations in phase from the ideal. In addition, frequency is related to phase error, such that the overall slope of the diagram indicates

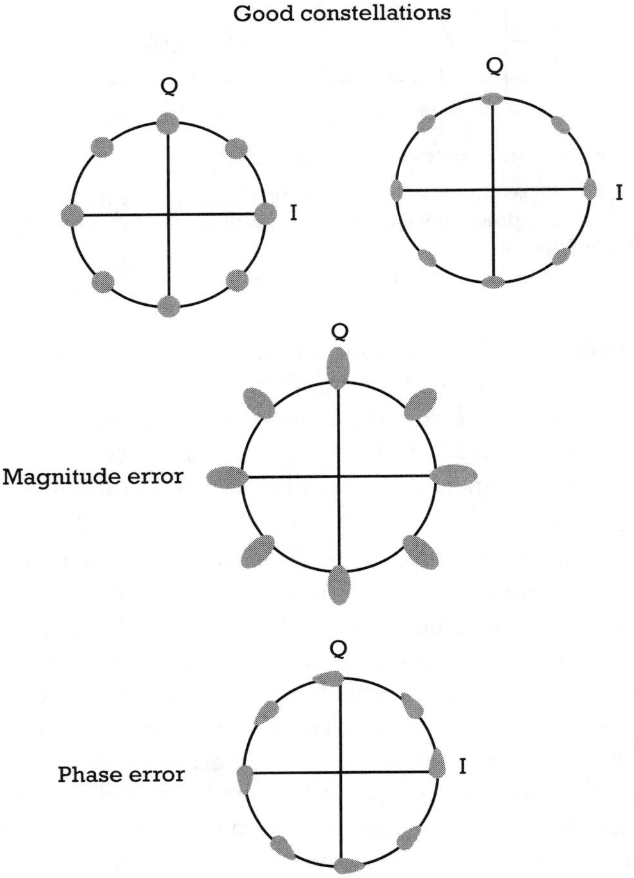

Good constellations

Q

I

Q

I

Q

Magnitude error

Q

Phase error

I

Figure 12.3 A constellation diagram can help quickly identify a modulation problem. Degradations in modulation due to amplitude or phase can be seen, particularly when the pattern of known good phones or BTSs can be used as a reference.

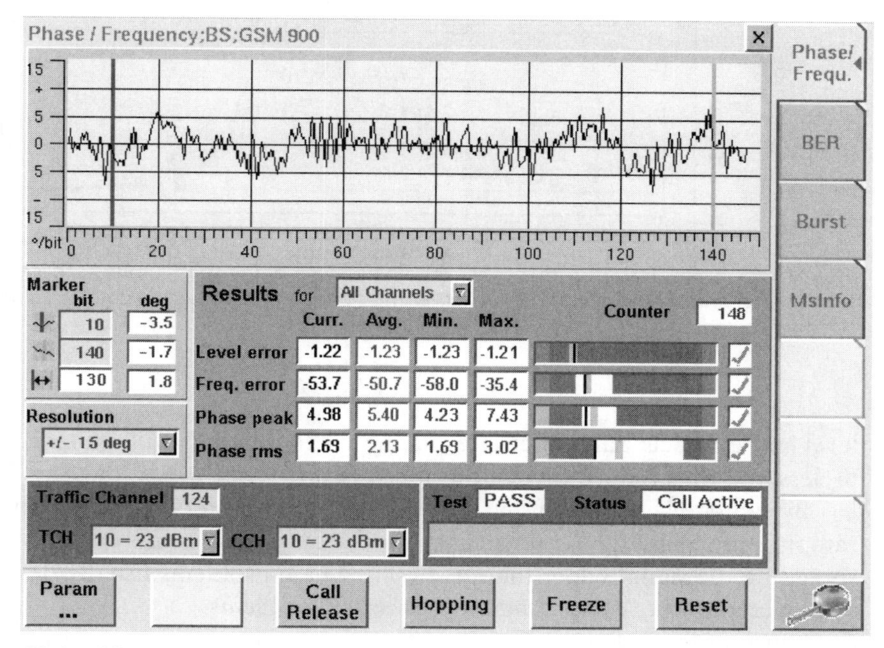

Figure 12.4 GSM test instruments will typically show a phase versus bits diagram. This graph can quickly show problems with modulation.

frequency error. Thus, this is a very quick way to judge the health of the transmitter.

- *Spectrum tests:* Two very important tests can be performed on base stations and mobile stations that use the frequency domain of a spectrum analyzer. The point of these tests is to ensure the TDMA signal has the correct occupied bandwidth, is not too wide, or has any unwanted spurious being transmitted into other frequencies.

In the research and development labs of phone manufacturers, engineers work diligently to ensure that the TDMA signal is modulated such that it will be within a specific channel spacing. In GSM, this is 200 kHz; in NA-TDMA, this is 30 kHz. If there are problems with the system, the signal could widen and begin to occupy more than the assigned frequency channel (see Figure 12.5). This will naturally cause problems in the network, as this signal will certainly cause interference to neighboring channels. There are typically two reasons for a signal to widen: problems with the modulator or transmit path (including excess noise) and problems due to switching.

In the first case, noise caused by modulation problems will increase the noise floor in adjacent and alternate channels (the channels next to the carrier). If this

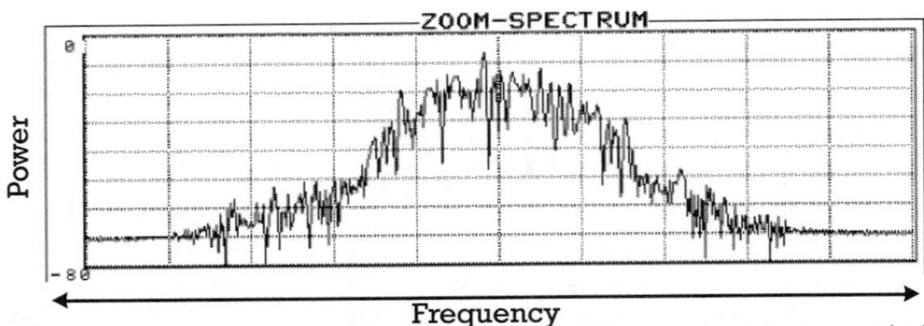

Figure 12.5 Most test instruments will automate some of the spectrum tests, optimizing the display so a technician can quickly see whether the signal is occupying the correct bandwidth or whether the adjacent channels are growing in power.

level in the adjacent and alternate channels is too high, it will render those channels useless and thus reduce capacity in the system.

By switching, we mean that the when the transmitter is ramping on and off, it is ramping too quickly. When a transmitter ramps too quickly, it has the effect of spreading the signal wider than it should. These problems usually develop because of power amplifier or leveling loop circuitry problems.

Spurious tests, described in Chapter 7, are fairly standard regardless of the technology. In TDMA systems, like others, it is important to make sure the transmitter is not emitting excessive power at any frequencies outside of the carrier. This is for government regulations, as well as for performance. If a transmitter is sending a signal somewhere else in the carrier's band, this will cause interference and, of course, limit capacity.

12.6 Receiver Tests

In order to understand receiver testing in digital systems, it is important to understand the setup. Because digital communications involves transmitting ones and zeros, the ability of a receiver to receive and demodulate, correctly determining whether the bit transmitted was a one or zero, is of paramount importance. The ability of the receiver to perform this task under difficult circumstances is equally important, as these are the circumstances the phone will run into with the customer.

These circumstances include low received levels, which naturally occur at the edges of coverage, in buildings or behind structures or in the inevitable coverage holes that will exist in any network. The difficult circumstances also include reception while driving at significant speeds or in conditions where there is significant multipath interference. Finally, the circumstances a phone will face will include situations in which there is significant cochannel interference because of system layout or in which all of the channels are being used in a high-capacity system, including the adjacent channel and several other channels around the carrier.

All of these circumstances are common and well known to the informed wireless technician, but they are completely invisible to the customer, who just wants to make a clear call to his wife while he drives down Fifth Avenue at rush hour. Hence, the phone and BTSs need to work in these harsh situations, and technicians need to perform the various receiver tests to make sure of it.

The receiver tests include:

- *GSM mobile station receiver test:* As mentioned earlier, the primary measurement in receiver test is BER. You will remember from Chapter 5 that the transmitted bits in GSM are encoded differently, depending on their importance to voice quality. Class 1a bits have error-correction encoding in addition to error-detection bits. Class 1b bits are error correction encoded, and class II bits are transmitted without either. This makes receiver testing a bit complex in GSM.

  For the most part, the standard field test will be on the class II bits, those bits with the least error encoding and protection. This is much simpler and quicker to implement, as it does not involve the error-correction process. For field purposes, this Class II BER is sufficient, as sensitivity is really a test of the front end of the receiver; this will affect all bits equally. If there is a need to perform more detailed tests, some test instruments do allow BER tests of the class 1b bits. Because of the error-correction process, this test generally takes longer, and, as mentioned, it is not always needed.

  In GSM phase 2, there is also a test mechanism in which the bits are looped back without any error correction. This loop is called the C loop, and has at least two advantages. First, all of the bits in the burst can be used for the BER (not only the class II or 1b bits), so it takes much less time to complete the measurement. Second, all of the bits are looped back as they are received in the mobile, without trying to correct them; thus, the true receiver sensitivity is tested without the effects of the channel decoder. Because phase 2 is fairly recent, most test sets do not support this method, although new platforms that do are being shipped.

  Finally, GSM phones also can indicate the number of bad frames it receives, based on the error-detection bits placed on the class 1a bits. The mobile phone will essentially "erase" the class 1a bits in frames that it determines to have too many bit errors. A frame erasure rate can then be determined to see how many bad frames were received compared to the number of total frames. If you understand that the mobile phone will erase frames when it determines errors on the class 1a bits, then you should understand that the class 1b and class II BER excludes frames that have been erased. This is why the class 1b and class II sensitivity measurement uses the term *residual BERs*, as it is actually the BER of the frames that have not been erased. Thus, it is important to examine BER, in addition to frame erasure rate in GSM handsets.

Thus, while SINAD is the basis for receiver test in AMPS, and BER is the basis in NA-TDMA, there are three bases in GSM: RBER Class 1b, RBER Class II, and frame erasure rate. You'll notice in the next section that the BER test can be performed by the test instrument by putting the phone in a loopback mode, while frame erasure rate is actually measured by the phone and reported back to the test instrument.

- *Setup for BER tests on mobiles:* There are basically two ways BER can be performed: the loopback method or the reported method. Because the reported method means the mobile phone will make the measurements based on error-detection algorithms, this method is quick. However, it is not as accurate nor as desired as the loopback method, as it assumes the phone is working correctly in the first place.

  In the loopback mode, the mobile station is placed into a test mode, where the output of the receive path channel decoder is looped directly back into the transmit channel encoder in the mobile station, bypassing the audio sections. In this way, the data bits are retransmitted back to the test instrument exactly as they are received. Because the test instrument knows what it originally transmitted (a PN sequence), it can compare the looped-back data and calculate a BER (see Figure 12.6). In NA-TDMA phones, the test instrument can send a control message to put the phone into this mode. In GSM phones, this requires a special SIM card, called a test SIM, to be inserted into the phone. Do not confuse this loopback mode with the audio loopback mode. In this case, the loopback is in the mobile station; in the audio loopback, it is the test instrument looping the data back to the mobile station.

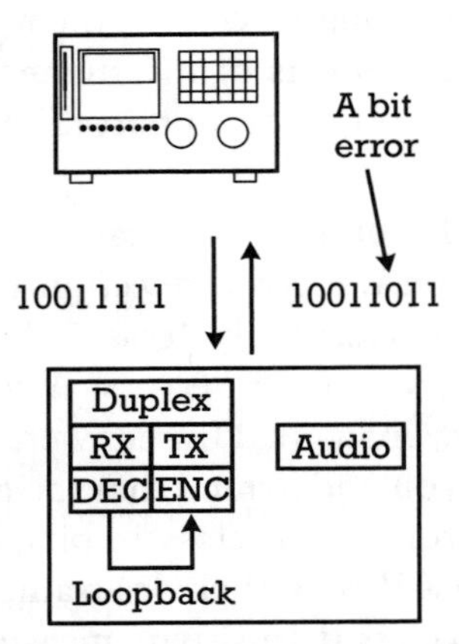

Figure 12.6 In the loopback mode, the test set transmits a PN sequence that is looped back by the mobile. The test set can then compare the transmitted sequence to the received sequence.

- *BTS BER test setup:* BTS BER testing is performed differently. Typically, the test instrument can generate a PN sequence in the reverse path. Usually, the test instrument assumes local control of the BTS, so it can either receive the bits back from the BTS and perform a comparison, or it can tell the BTS the pattern and have the BTS make the comparison or measurement.
- *Sensitivity:* In TDMA systems, sensitivity will be the lowest level a system can receive a signal and still have a useable BER. Simply, the test setups are employed, then the RF level is lowered. Typically, there are two ways to perform the sensitivity test. First, a set of levels can each be set and a BER measurement made at each, assuring proper operation at the levels required. This is the typical test performed in the field. A second, more advanced, approach is to determine the absolute sensitivity by monitoring the BER while lowering the RF level until the BER exceeds a specific threshold.
- *Measurement reporting:* As mentioned earlier, the mobile stations will report received signal strengths, as well as making measurements of BER estimates. It is important that these systems are working correctly, as they are used extensively in the handoff process. Remember, TDMA phones can measure other channels while on a conversation, and this is how BTS selection during MAHO is accomplished. Thus, a typical measurement method for this function is to induce bit errors into the transmit stream, as well as vary the RF level of both the carrier and the "second" channel that a mobile station should be monitoring. Assuming the test instrument is accurate, the mobile station should respond accordingly. This type of test is not typically performed regularly on BTSs; however, it can be performed in a similar manner, where the test instrument will insert a known sequence with bit errors and at a certain level, and the BTS should report back the results accordingly (this type of testing requires a protocol analysis system on the Abis interface or an Abis interface simulator—the Abis interface is that between the BTS and the BSC).
- *Susceptibility:* There are four receiver tests that come under the heading of susceptibility tests. These are tests that test the system's ability to operate in environments where there are potentially interfering signals. Typically, these tests are not performed on BTSs in the field, although they are in laboratory and manufacturing environments with similar setups. These tests include cochannel rejection, adjacent channel rejection, intermodulation rejection, and spurious response. Sometimes you may hear these tests referred to as selectivity tests, in that they measure the mobile station's ability to select the desired signal in the presence of undesired interferers (see Figure 12.7).
- *Cochannel rejection:* This is a test of the mobile station's ability to receive a channel off of one BTS, when there is a signal from another base station reaching it in the same channel. Typically, during this test on a mobile, an interfering signal will be injected 9 dB down from the wanted signal at the same

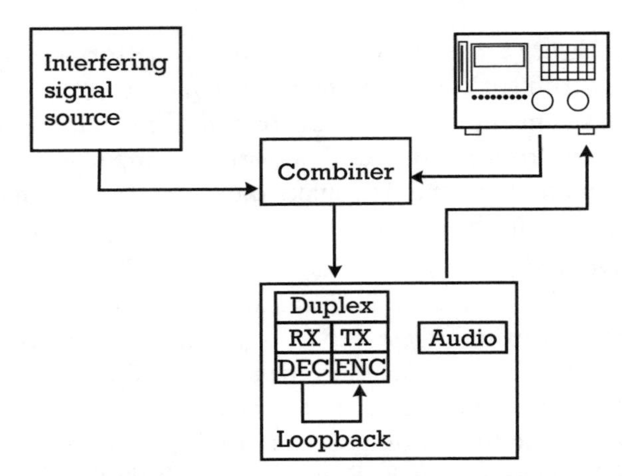

Figure 12.7 For the susceptibility tests, an interfering source is combined with the test set's signal before being input to the receiver. A standard BER test is then performed.

frequency, and the BER should still remain above a useable level. In today's networks, where tower space is at a premium, networks cannot always be designed as wanted, so some cochannel interference is inevitable, and a mobile station must be able to operate in this environment.

- *Adjacent channel rejection:* This is a test of the mobile station's ability to receive its assigned channel when the adjacent channel is being used and is at a significantly higher level as well. In areas of very high usage, this may well be the norm, so the mobile station must be able to select its channel and not be caused problems by a neighboring channel. In this test, a mobile station establishes a call while another channel 9 dB higher is applied in an adjacent channel. BER is what determines the performance.
- *Intermodulation rejection:* This is the ability of a mobile station to operate correctly when several signals at specific offsets from the carrier are applied to the receiver during a call. These specific offsets are designed to create intermodulation products within the wanted signal channel. Again, BER is measured after the interferers are added to ensure proper operation.
- *Blocking and spurious response:* These are similar to intermodulation rejection, except the interfering signals are anywhere in the band from 100 kHz to 12.5 GHz.
- *Demodulation in Doppler shift and multipath fading:* Doppler shifting occurs when a receiver attempts demodulation while traveling at high speeds. The effect is changes of the phase of the received signal, which will cause problems in the demodulation process. Multipath fading was explained in detail in Chapter 1. These receiver tests require a test system that can add/create Doppler shifting and multipath fading, and then measure the BER. The amount of

Doppler shifting is typically converted to a speed (e.g., a test might simulate the Doppler shift of a car moving at 55 miles per hour). This ability is fairly common in test sets aimed at field use. Adding multipath interference, however, generally requires very complex and expensive test equipment in addition to the communications test set and thus is generally only performed in research and development labs and manufacturing environments.

12.7 Audio Tests

As mentioned earlier, TDMA mobile stations use vocoders, so audio testing is a bit more complex than it might seem. For field test purposes, an audio loopback test is performed in which the technician can speak into the microphone, with the speech being transmitted to the test set. The test set then pauses a moment and then retransmits this data back to the mobile station, which can decode and play the audio back to the technician. If the voice sounds good, it can be assumed the vocoder and audio sections are working properly. More advanced objective tests on audio can be performed, but these are primarily for research and development and evaluation labs, and the tests are very complex and quite expensive to implement. They also require a test set with its own vocoder, so that the audio can be broken out and sent to voice analysis systems. For field purposes, the qualitative audio loopback test is sufficient.

12.8 Testing Data—GPRS and EDGE

You will recall from earlier chapters that GPRS uses the same RF structure as GSM. Hence, RF testing is identical to GSM. EDGE adds a new air interface, so here tests will be different; however, all of the concepts remain the same with regard to test concepts.

What is different with data formats is that the protocol stacks and application layers are significantly more complex. Whereas in voice communications, a technician needs to ensure the vocoder is working correctly, in data transmissions, several layers and interoperabilities need to be ensured are working correctly. For the most part, to a technician this simply involves a go/no go test to ensure the data service is working properly.

In addition, several RF tests might become more important in data than in voice, as the impact or complexity of the design is different. For instance, because of some of the changes in the way GPRS is implemented, timing in GPRS is a test that needs to be thoroughly checked.

In terms of network testing, often times an end-to-end test suffices, as it will ensure that all layers, from RF to application, are correctly operating. That said, several metrics can be looked at to ensure customers are getting a quality service.

For instance, data throughput, data call setup time, and packet retransmissions can be examined. The problem with testing these metrics is poor quality can be the result of not only the RF, but the backhaul and beyond side. Hence, they are excellent tools to ensure end-to-end service is working, but for testing actual RF, it is better to rely on specific RF tests discussed earlier.

12.9 Conclusion

TDMA systems have several new test methods in order to assure proper operation when used by the customer, but the general concepts remain the same. Many technicians lose track of the fact that even with digital encoding and complex transceivers, the digital cellular system is still a radio system. Testing data systems, for an RF technician, will not be very different than voice. For the most part, RF is RF, regardless of whether data or voice is being transmitted.

CHAPTER 13

Testing CDMA

13.1 Introduction

When it comes to testing, many engineers would agree that CDMA systems offer the most complex challenges. With the many unique concepts involved in CDMA, it often becomes difficult for test engineers and technicians to decide which tests are really required in the field and which tests are truly design tests, needed only in design stages. With a solid understanding of the major technological concepts, however, a technician or engineer can easily see the similarities between the suite of tests performed in other technologies and those performed in CDMA. Once again, there are tremendous differences in how certain tests are performed, but most of the tests in CDMA have counterparts in AMPS and TDMA systems.

In Chapter 1, this text spent a great deal of time discussing domains. In CDMA, a new domain is added: the code domain. Of course, when we talk about specific tests, the domain of the measurement can be very important. For instance, when we discuss spectrum analyzer measurements, we are using a frequency domain. When we measure the burst of a TDMA signal, we use the time domain. In CDMA, we will make measurements that take into account these two domains, frequency and time, in addition to the third domain, the code domain. You should remember that CDMA divides users with unique codes, bursts the power in the uplink, and transmits its 1.2288-MHz signal at a specific frequency. In addition, CDMA's complex modulation and spreading process makes modulation testing a bit different, although constellation analysis can still be used to test the QPSK modulator. To account for the more complex modulator and spreading, a new measurement, RHO, is tested. CDMA also has a very complex power control system. You should recall that this power control system's correct operation is vital to optimized performance of a CDMA network as a whole, and thus proper testing is very important.

What is important to understand is that despite the complexities of CDMA technologies, it is still essentially a radio and thus contains the same basic components of a traditional radio that requires testing.

13.2 Introduction to the CDMA Transceiver

The CDMA transceiver is a bit more complex than analog and TDMA transceivers, to be certain. This is because in CDMA systems, an additional process is added. Both TDMA and CDMA systems include voice, channel, error-correction and error-detection encoding and decoding, interleaving, and of course final RF modulation for transmitting. CDMA, however, adds various spreading processes in order to create the spread-spectrum signal. Additionally, CDMA uses a much more complex control and receiver subsystem.

Even with the increased complexity of CDMA systems, for the most part, these complex subsystems are incorporated into chips and small packages, which generally are not field repairable for the average technician in the field. Thus, technicians generally need to have an understanding of how CDMA works in order to isolate a particular module, but detailed understanding of components is not required (this is really the same for all systems built in recent times).

The block diagram shown in Figure 13.1 is a general representation of the many processes involved in CDMA. Of course, each manufacturer will have their own specific block diagram, which could vary greatly from the one shown, as it is meant to be very generic in its representation. The processes shown have been discussed in detail in the CDMA technology chapter.

13.3 Introduction to CDMA Tests

The CDMA BTS and mobile station have significant differences in operation. For one thing, the CDMA BTS transmits multiple code channels simultaneously, while a mobile station will only transmit one channel. The mobile station bursts its power as it changes its vocoder rates. The mobile station also uses a very complex power control system. In addition, the mobile station compacts call processing, memory, special "PCS" functions, and user interfaces into its package, all of which need tested in the field. The BTS does not need such extensive testing, as most of this control comes from the BSC. Often a phone can be used to verify performance, while it would be impractical to use a real BTS to test mobile stations.

Because of the significant differences between test procedures compared to TDMA and AMPS, and the differences between BTS and mobile station testing, this chapter will have a slightly different format than the previous two chapters.

CDMA requires precise timing synchronization in order to operate correctly. It is for this reason that every BTS in a CDMA system has its own GPS receiver, which, of course, gives every BTS in the network the same exact timing reference. The BTS then sends this timing information to each mobile station via the pilot and sync channels. It thus makes sense that if a test instrument is to test this BTS signal, it will also require this precise timing. For this reason, BTS test sets should either include a

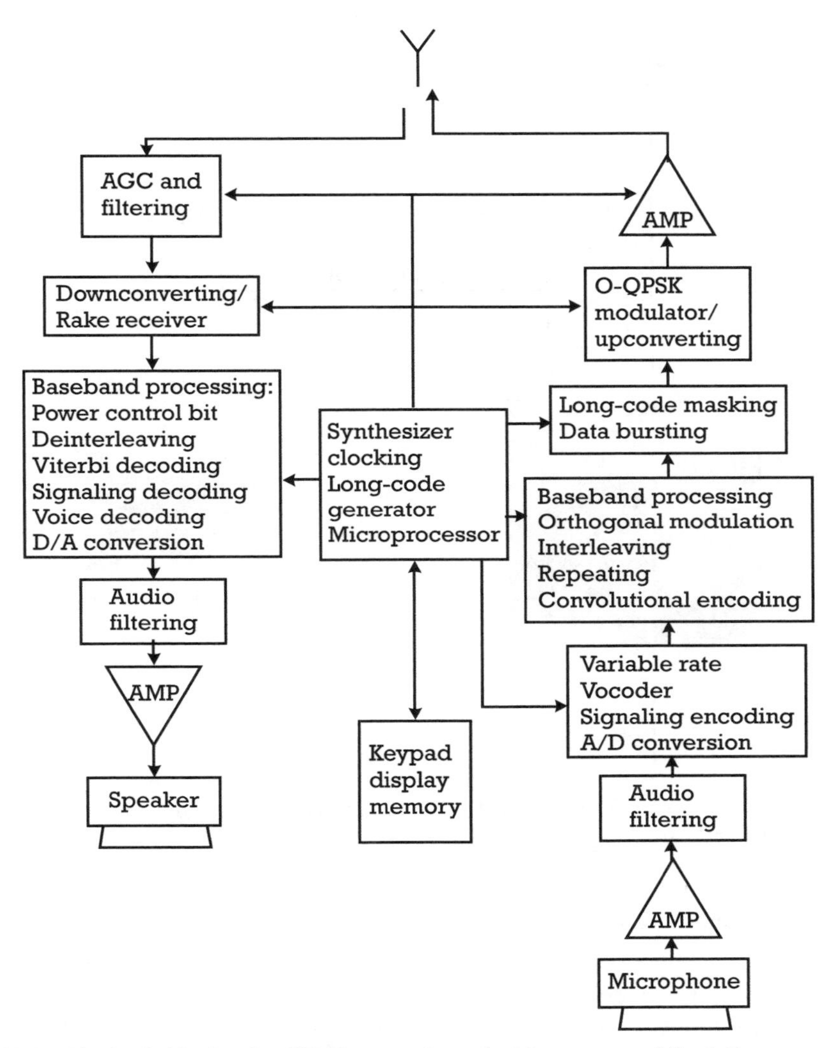

Figure 13.1 The basic blocks of a CDMA transceiver—in this case, a mobile station.

GPS receiver of their own or allow for the input of the timing signals from the BTS (see Figure 13.2). Typically, the signals required for input include the even-second clock, a derivative of the GPS one-pulse-per-second clock. As such CDMA channels as the pilot and sync channels are triggered by this pulse, this will allow the test instrument to trigger as well. It also serves as a reference for the test set to make timing measurements. A second frequency reference is also required in order to make accurate measurements. Typically, this is a 10-MHz signal, which again is taken from the GPS signal. Alternatively, some BTS manufacturers use a derivative of the

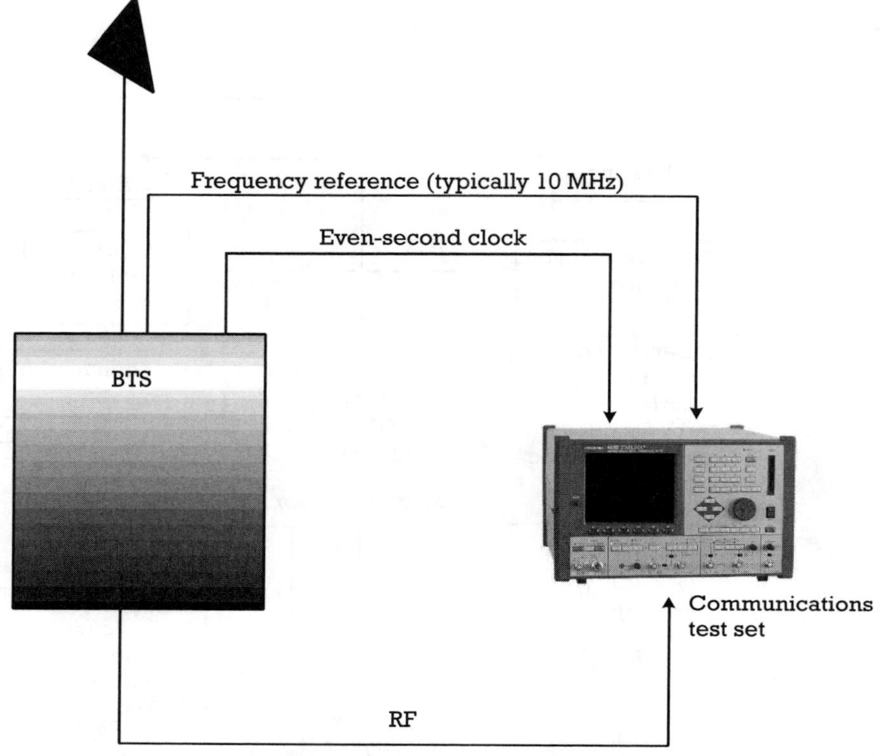

Figure 13.2 In order to test CDMA base stations accurately, timing and frequency signals are needed from either an independent GPS receiver or the BTS itself.

CDMA chip clock, and most CDMA BTS test sets can automatically detect which clock is used and adjust accordingly. Regardless of the reference frequency, it is, along with the even second pulse, required for test sets to measure a BTS correctly, whether they are taken from the BTS itself or an independent GPS receiver as part of the communications test set.

When testing mobile stations in the field, such timing signals are not required. Remember, the mobile station receives its timing from the BTS over the air interface; thus, the test set must also send this timing information during the testing.

13.4 Testing Power

Testing power in a CDMA system can be much more difficult than other systems. You will recall from Chapter 7 that the preferred method of measuring power in CDMA is via the channel power measurement. In this way, the average power only in the transmitted channel is measured. For mobile stations, power testing gets very complex, as the power control operation of the mobile needs to be thoroughly tested. Power problems in CDMA usually point at the amplifier section; however, the problem can be caused anywhere in the transmit path after modulation. In mobile stations, because the mobile must correctly demodulate and process the power control information before it can react to it, power problems can be difficult to isolate, although a good understanding of the test procedures should help narrow down such problems.

Power tests include the following:

- *Channel power:* The channel power measurement is very important in BTS testing, particularly during deployment, as the channel power that will be set will be the principal initial determinant of cell size. You should recall from Chapter 6 that there are many other determinants in CDMA, such as uncorrelated interference, code domain power ratios, and angle of the antennas. Of course, cell size is not the only important reason to accurately measure channel power. Too much channel power may cause too much interference in a neighboring cell, which can lead to capacity problems.

  Remember, interference is intrinsic in CDMA; after all, all of the BTSs are transmitting on the same frequency. With the Rake receiver, this is not always a bad thing, as the mobile station can receive several BTS signals simultaneously and combine them coherently during a soft handoff. Problems arise when there are too many signals reaching a mobile station. Any BTS signal that is not being demodulated by a mobile station's Rake receiver is noise to the receiver. Typically, if there are more than four BTS signals above a usable level reaching a mobile station, performance will begin to get degraded, as the noise floor will be raised. This situation is called pilot pollution, as too many pilot signals are present at a location and are polluting the system (see Figure 13.3). Pilot pollution generally causes decreased capacity in a particular area, which to your customers equates to blocked calls—or, worse yet, dropped calls as the user tries to hand off into the area.

  The idea is to limit the bad interference as much as possible, and this is realized by proper network layout in addition to regular and accurate power calibrations of BTSs. While pilot pollution is not typically a problem for technicians, it is important for performance engineers and mobile station technicians to understand the concept, as it is can be a common cause of problems. Low-cost drive test systems that can monitor the power in received pilots are often given to BTS technicians so they can quickly determine if a called in complaint is perhaps caused by pilot pollution or if it is indeed a hardware problem with the BTS.

  Channel power is typically measured with the pilot channel only. While some manufacturers do recommend a broadband power meter to make these measurements, this is impractical if multiple carriers are used. It is also

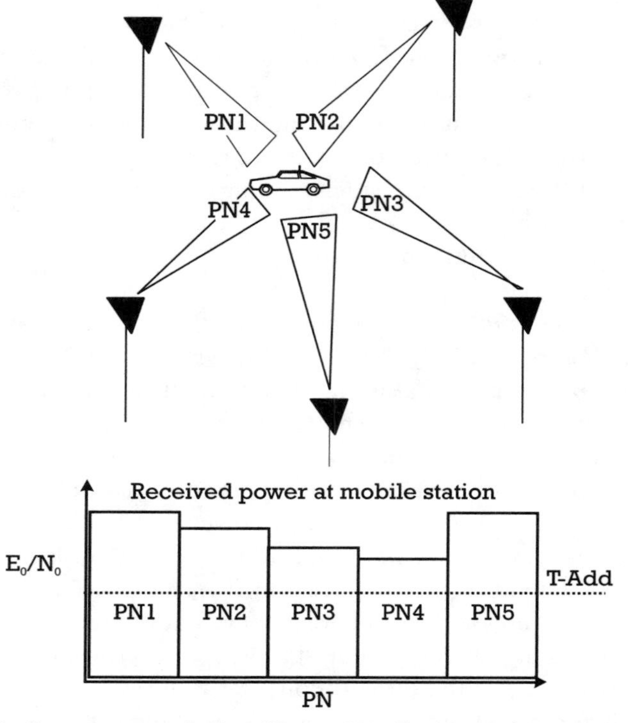

Figure 13.3 In this diagram, power in five different PN offsets is reaching a particular location. Because too many (more than four) of the PN offsets are above the threshold needed to be active (T-add), this can cause a pilot pollution problem.

impractical if there is a problem with the BTS causing excessive out of band energy. Both of these scenarios would result in higher than correct values. The values measured are typically greater than 20 dBm, which will often require that an attenuator be placed between the BTS and the test instrument in order to protect the test set's front end from damage due to excessive input power. Some newer test sets allow for very high power levels to be input and thus do not require the clumsy attenuators be placed inline.

When it comes to mobile stations, perhaps no other testing is as important, other than basic call processing, as power-control tests. The two forms of reverse-channel power control in a CDMA system, closed and open loop, must be thoroughly tested, even in the field. Also, mobile stations use access probes to get service; these probes need to be tested to ensure a mobile station can perform access functions optimally. It is important to understand that these power tests for the mobile stations are both power-accuracy measurements, as well as functional tests, ensuring the mobile station can correctly identify and react to the power control situation.

- *Open-loop power control test:* This is performed by the test station sending alternating ones and zeros as the power control bits. This ensures that the closed-loop power control will not cause any varying in the output power of the mobile station. At this point, the test set (which is simulating a BTS) will vary its own output power in the forward link. The mobile station's open-loop power control, you will recall from Chapter 6, is dependent on the amount of power it receives from the BTS. Thus, if the mobile station receives less power, it would simulate the mobile station getting further from the BTS, and thus the open loop power control should react by increasing the output of the mobile station accordingly. The test is not so much that the mobile station reacts, but that it reacts quickly. Thus, this test is best performed using a graphical representation in the time domain. The test set varies its output power for 100 ms, and the mobile station should vary its power within a specific mask, shown in Figure 13.4.
- *Access probe power tests:* Access probe tests must also must be performed, as this is the method by which the mobile station will try to gain access or respond to a BTS. The parameters that need to be verified include nominal power offset, initial power offset, power increment between consecutive probes, number of access probes in one sequence, and the number of probe sequences in one access attempt (see Figure 13.5).

  Essentially, this ensures the access probes are occurring at the correct power levels, that the mobile station is transmitting the correct number of probes, and that the power is being increased correctly. Remember, a mobile station will send an access probe and await a response from the BTS on the paging channel. If it does not get that response, it will send another probe, this time at a higher level. If this process is not working correctly, a caller will occasionally get blocked. The parameters listed here are all settable by the network via the access parameter message.

  If a mobile station is not sending all of the access probes, or is not increasing each successive probe correctly, this could cause a very aggravating condition for a customer, who might be able to make calls occasionally, but might get blocked from making a call sporadically. Worse yet, a customer might miss calls because registration attempts failed. This could be caused by a faltering amplifier or the control circuitry associated with the access probes.
- *Closed-loop power range:* The range of the closed-loop power control must also be tested. In this test, the test set will transmit a constant power level, thus ensuring no variations from the open-loop control. The test set will send 100 consecutive zeros and ones as the power control bit and measure the range and functionality of the closed-loop process. This is done at all of the data rates. The range of the closed-loop power control must be ±24 dB around the open-loop estimate.

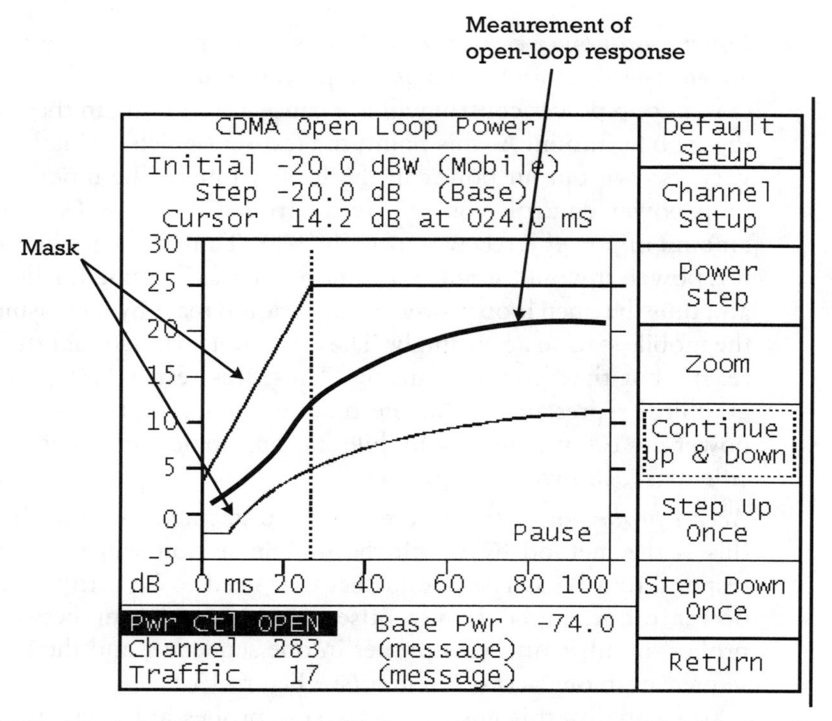

Figure 13.4 A typical display of a test set during an open-loop power test. Note that the measured power must stay within the confines of the mask. The power is measured across a 100-ms window (the time domain).

- *Minimum and maximum power:* Two further tests that combine both the operation of the closed- and open-loop system are the maximum and minimum RF output power tests. For the minimum test, a –104-dBm base station signal is sent, thus maximizing the open-loop power control. Having all zeros on the power control bit to ensure the closed loop is at its ceiling. The specification will vary depending on the mobile station class from 1W to 6.3W. The minimum output power test is essentially the same. The maximum received BTS level is sent (–25 dBm) with all ones on the power control bit. The mobile should then be transmitting less than –50 dBm. The maximum power test is important, because this, along with receiver sensitivity, will limit how far away from a BTS a mobile station can be and still have adequate service. Minimum power is important from a network perspective. The mobile should be able to generate low levels when it gets close to a BTS, so as not to generate excessive interference in the network (which, again, lowers capacity).
- *Gated output power:* Another power test for mobiles is similar to burst power versus time tests performed in TDMA. Gated output power is a measurement

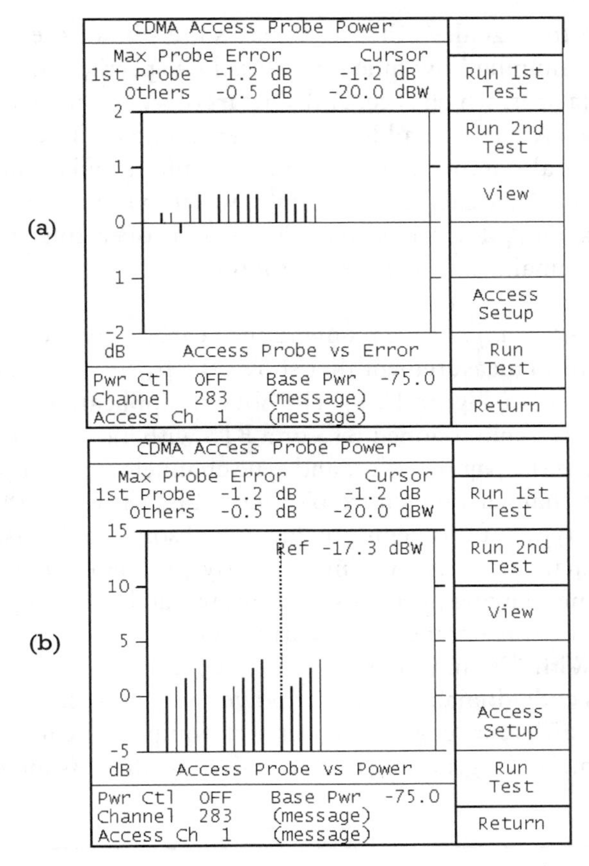

Figure 13.5 Two tests are typically run by test sets to test the access probes. (a) The first test shows the access probes being increased in power, as the mobile does not get a response from the BTS. (b) The second test ensures the probes are accurate in power.

of the time response of a mobile during variable data-rate transmissions, when the mobile is only transmitting during gated-on periods. This time response must fit in a given envelope mask, similar to TDMA. In CDMA, the reverse link signal will burst in 1.25-ms windows, at a rate dependent on voice activity.

13.5 Code-Domain Tests (BTS Only)

Each BTS has the ability to transmit on any of the 64 Walsh channels simultaneously and combine them. If you were to look at a spectrum analyzer's display of a CDMA signal, you would see one bart-head. This doesn't tell you a lot about the makeup of the signal. A CDMA signal is the sum of the power in all of the Walsh codes. Determining how much power is in a specific code channel requires a code-domain analyzer, which is generally a part of a communications test set designed for testing CDMA BTSs. In addition to determining the fraction of power in each code channel, it is also important to examine the phase and timing of the individual channels to ensure that each code channel, which will be used for individual conversations or paging, pilot, and synchronization, is operating correctly.

Code-domain tests include the following:

- *Code-domain power:* As mentioned earlier, code-domain power is generally a graphical measurement that shows the fraction of power present in each code channel (see Figure 13.6). The obvious application of this test is to ensure that a newly commissioned BTS or a BTS with a reported problem is indeed transmitting the paging, pilot, and sync channels at the appropriate levels. Without these channels, the entire BTS would be useless to the network. Quite often, these types of problems are due to bad software loads or troublesome individual channel cards. A common early problem was with the synchronization channels having problems, and this would be the only way to detect if the sync channel was on and at the correct level.

  With 1X and later variations come variable-length codes. This complicates code-domain testing, as power will be spread across different codes that have different lengths. Newer test equipment can simplify these tests for a technician by grouping related codes together. As mentioned earlier, however,

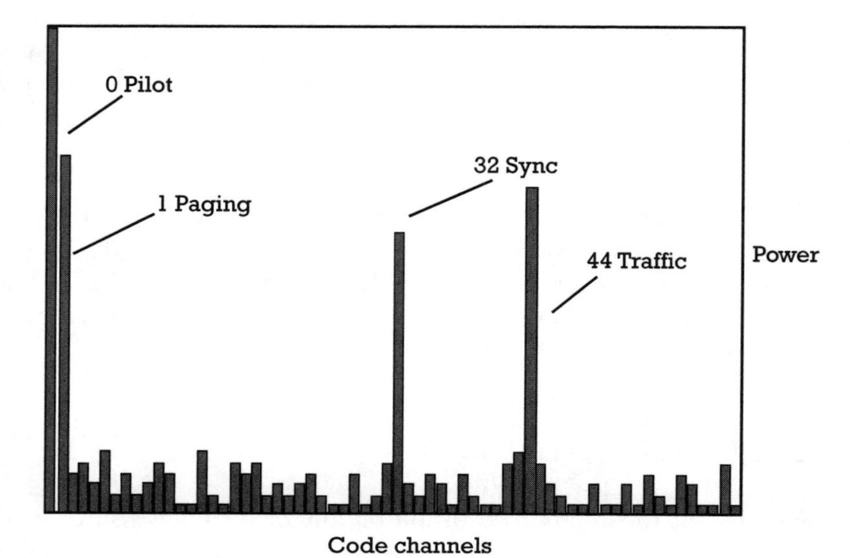

Figure 13.6 A typical code-domain analyzer screen.

often it is the codes that should have no power that can quickly indicate a problem, and this remains true in CDMA2000.

A less obvious, but more important, code-domain test does not have to do with the active channels but rather the inactive channels. The 64 code channels in each carrier will show all of the power being transmitted in a particular channel, whether it is correlated with a particular code channel or not. Any error power will thus show itself as a rise in the inactive channels. Most commonly, error power is completely uncorrelated with any particular code channel (see Figure 13.7), and in this case, the power will spread itself evenly across all of the code channels. This raises the code-domain noise floor, which will cause a decrease in the capacity of that BTS.

Occasionally, a problem can cause correlated error to appear in particular code channels. This is generally due to the concept that code channels act like sine waves, in that they will mix when passed through nonlinear devices. For instance, code channel 1 might mix with code channel 32 to create a correlated error power problem in code channel 33, as in Figure 13.7(b). Code channel power problems such as these are typically caused by intermodulation in the transmit path causing inband interference; thus, they can be caused by a bad amplifier or component in the transmit path, as well as by the modulator.

- *Code-domain timing:* What separates the code channels is their unique code sequence used to spread the data, which make each channel orthogonal to the next. If the code channels do not all have the same timing, then they will lose this orthogonality (they are all orthogonal assuming they are time aligned). If code channels lose their orthogonality, they can interfere with each other, as they will not be perfectly uncorrelated to each other. The code-domain timing graph will show the timing offset of each code channel, relative to the pilot channel. If a code channel is more than 50 ns off of the pilot channel, there could be interference. Typically, this is an adjustable parameter via software, although intermodulation or problems in the modulator can cause such problems.
- *Code-domain phase:* Mobile stations get their phase alignment from the pilot channel. When they switch to a traffic channel for a conversation, the mobile station still uses this original alignment for its phase reference as it demodulates the traffic channel. Thus, if a BTS is transmitting on a particular traffic channel whose phase is more than 50 mrad different from the pilot channel, the mobile station will have difficulty demodulating once on the call and could drop calls. Phase problems can be caused by problems in the modulator section of the BTS, as well as intermodulation between the code channels.
- *Pilot power:* While not specifically a code-domain test in the sense that it does not use a code-domain analyzer, pilot power is the amount of power in the pilot channel, relative to channel power. Pilot power, measured in decibels

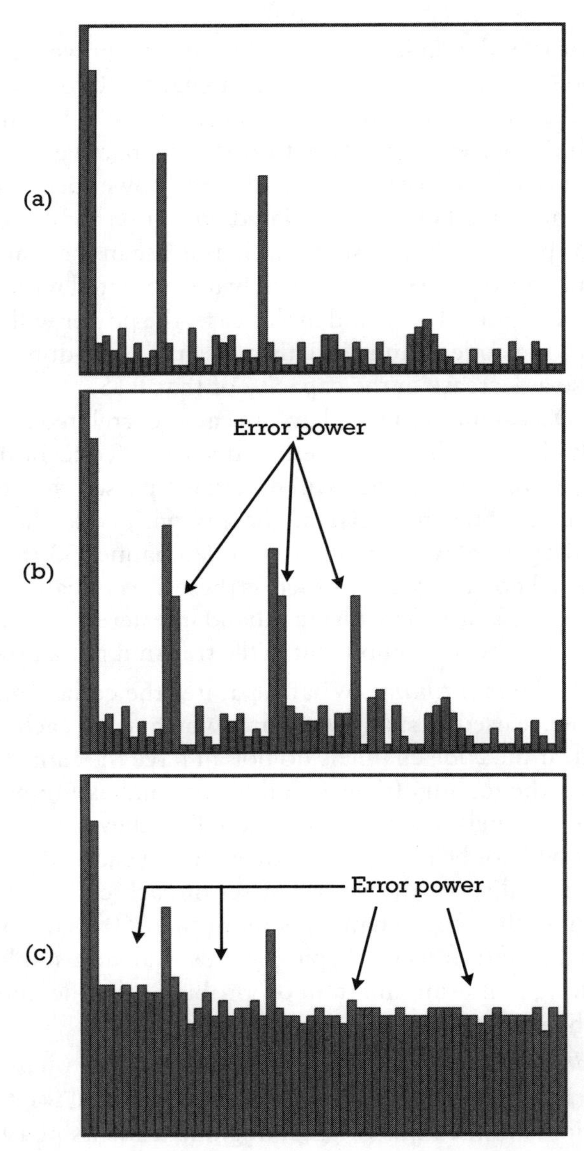

Figure 13.7 The code-domain analyzer can show problems in the CDMA BTS transmitter: (a) ideal transmission, (b) mixing caused error power in specific codes, and (c) uncorrelated excess power in the transmitter band will spread all codes evenly, causing a rise in the noise floor.

below total power, is set to a particular level by the manufacturer and/or carrier and should be verified during routine maintenance. Deviations from this setting can lead to performance problems in the network.

13.6 Timing, Frequency, and Modulation Testing

Of course, timing and frequency are very important in a CDMA network. Exact synchronization is absolutely needed to ensure that the many sequences used in the system are received the same way they are transmitted. Modulation needs to be tested to make sure voice and signaling quality remains high. If timing, frequency, and modulation problems exist in a mobile station or BTS, it can lead to poor access and dropped calls, bad voice quality, and excess error power being transmitted. This decreases the overall S/N ratio, which, of course, leads to decreased capacity. If the EVM is high, this will cause error power to be transmitted in the system (power for a bit that is an error bit becomes error power).

Timing, frequency, and modulation testing includes:

- *Time offset:* CDMA is based on absolute timing signals received from either a GPS system in the case of the BTS or timing relayed from the BTS in the case of the mobile station. If the timing is off on either the mobile station or BTS, a call will not be able to be established.

  In the base station, this timing offset measurement is performed on the pilot channel, as it is the primary timing reference for the system. In the BTS, all of the other code channels are based on the pilot timing. You should remember that the pilot channel is triggered by the even-second clock. Thus, in order to perform this measurement, the test instrument will need to have the even second directly as a reference. The test set can then determine the amount of time between where the pilot sequence should have begun and where it actually began. In BTSs, it is very important that this specification be met. This is because if one BTS has a high time offset compared to all of the surrounding BTSs, a mobile station will be unable to synchronize with this base station as it moves into the cell, and the call will be dropped. Note that a bad time offset will not affect mobile stations already in the cell, as mobile stations derive their timing from the BTS.

  The mobile station receives its timing from the BTS. If the mobile station cannot properly align itself to the timing of the network, it will be unable to properly initiate service on the network. In the same way as stated earlier, the test set determines the timing of the mobile station and how much it differs from where it should be.
- *Frequency offset:* Frequency offset is important for obvious reasons. The mobile station and the base station must have the same frequency in order to operate correctly. The BTS derives its frequency from the GPS receiver, so there are actually two frequency tests—one to test whether the GPS receiver is locked correctly, and one to test whether the BTS has locked correctly on the GPS reference and is not drifting. Mobile stations need to have their frequency

offset measured as well. In addition, if frequency is off, it can also cause interference with other channels.

- *EVM and carrier feedthrough:* These measurements were discussed in previous chapters and are very similar. The constellation diagram should still be used to determine how well the modulator is working (see Figure 13.8). See the previous chapters on how to use the constellation diagram, which many test instruments will have.
- *Waveform quality or RHO:* In CDMA, what is most important is limiting error power. In the field, a primary cause of error power will be problems with modulation, although there can be other problems as well. While code-domain measurements can be used to determine some error power problems, RHO can help determine if there is error power in band and particularly whether this error power is due to modulation. RHO is a correlation measurement. The test set must know what is being transmitted in order to compare what this signal should ideally look like and what it actually looks like. For BTS test, a

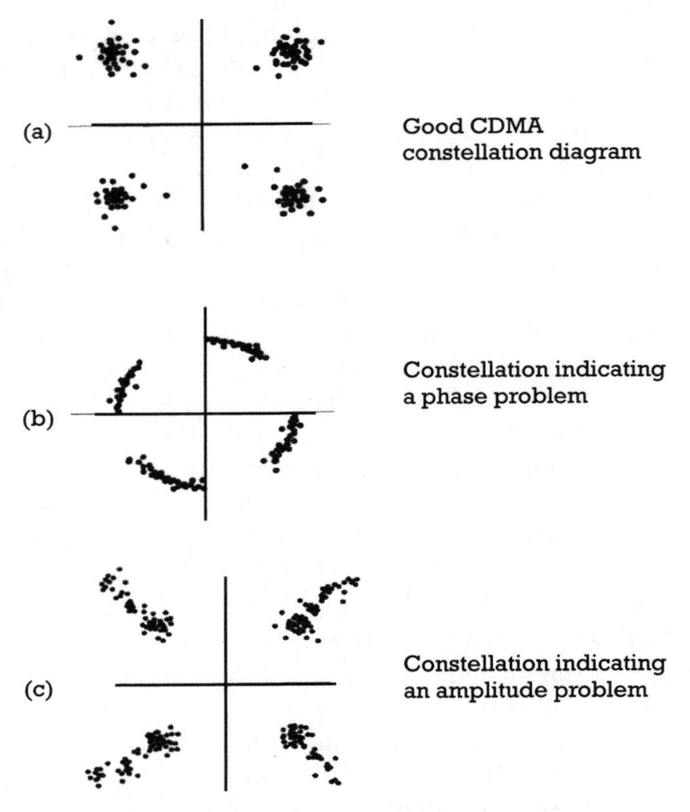

Figure 13.8 (a–c) Constellation diagrams for CDMA signals. This is a quick way to spot a modulator or transmit path problem.

pilot-only configuration is used for the test. In mobile stations, a loopback mode is generally used.

As in all correlation measurements, a one means perfect correlation, while a zero means no correlation. Thus, RHO is a general indicator of overall modulator performance. RHO, therefore, is a general indicator of a problem somewhere in the system, and thus more testing should be made to isolate the exact cause. If RHO goes down either on base stations or mobile stations, this means more error power is transmitted, which decreases capacity. For mobile stations, RHO must be above 0.944; for BTSs, RHO should be above 0.912. This relates to 94.4% correlated power for BTSs and 91.2% correlated power for mobile stations.

Problems with any of these tests can have a variety of causes. Intermodulation, modulator problems, frequency locking circuitry, and spurious all can cause problems with these measurements. If RHO is bad, typically a technician can take a look at a constellation diagram, which can quickly show if there is a problem with phase, magnitude, or perhaps carrier feed through. If this constellation diagram looks good, the problem may be with uncorrelated power in band. At this point, the code-domain power analyzer can be used to determine if this might be the case.

13.7 Call-Processing Tests

Like the other formats, the various call-processing functions of mobile stations need to be tested to ensure functionality. These include mobile origination, mobile station page, authentication, short messaging, caller ID, and other such PCS features. Descriptions of these tests can be found in Chapter 9.

In addition to these tests, which are mostly common to all technologies, CDMA has a complex handoff subsystem that needs to be checked. While performing many of the advanced complex soft-handoff tests would require multiple base station simulators, most CDMA mobile station test sets allow for basic functionality testing of the softer handoff. While there is no formal softer handoff test for the field, ensuring the mobile station is able to perform a softer handoff is really all that is needed for field purposes, much the same way hard handoffs are tested in other formats. Of course, CDMA has hard handoffs as well, so these will need to be tested as well in a similar manner.

13.8 Out-of-Band Tests

CDMA still requires a spectrum analyzer to make sure no excess power is being transmitted out of the transmit band. This not only can indicate a possible problem with the transmitter, but it also is often regulated by governmental agencies (like the

FCC). In the field, these tests are not tremendously important because filters have been used to limit these out-of-band emissions, and these filters seldom degrade. It can be useful to look at the spectrum of the CDMA signal to make sure the slope of the signal is vertical enough as well as to ensure the *shoulders* are down far enough relative to the channel power. If the slope is not vertical enough, this could indicate a filtering problem. If the shoulders are too high, this could be the indication of an intermodulation problem somewhere in the transmit path.

13.9 Receiver Tests

CDMA receiver tests are similar to the receiver tests performed for other systems (see Figure 13.9). Sensitivity, dynamic range, and several susceptibility tests that test the ability to receive under different conditions with various interfering signals are all tested in CDMA.

Because of the very complex error-correction system used in CDMA, typically either all of the bits in a frame will be perfect or many of the bits will have errors, so BER is not really a good indicator. In CDMA, FER is used. FER is the ratio between frames with errors versus the total number of frames transmitted.

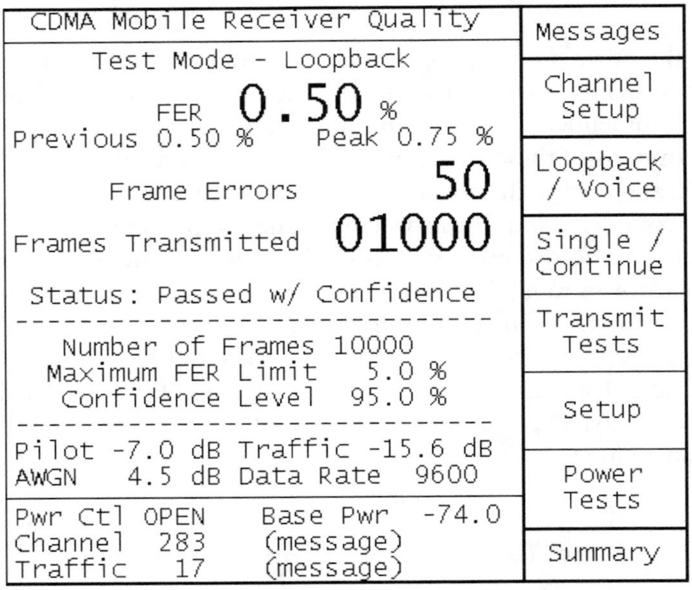

```
CDMA Mobile Receiver Quality          | Messages
Test Mode - Loopback                  | Channel Setup
FER 0.50 %                            | Loopback / Voice
Previous 0.50 %    Peak 0.75 %        | Single / Continue
Frame Errors                 50       | Transmit Tests
Frames Transmitted        01000       | Setup
Status: Passed w/ Confidence          | Power Tests
-------------------------------       | Summary
Number of Frames 10000
Maximum FER Limit    5.0 %
Confidence Level    95.0 %
-------------------------------
Pilot -7.0 dB Traffic -15.6 dB
AWGN   4.5 dB Data Rate   9600
Pwr Ctl OPEN    Base Pwr   -74.0
Channel  283    (message)
Traffic   17    (message)
```

Figure 13.9 A typical mobile station receiver test report on a test set. Note that the screen shows number of frames, number of errors, the computed FER, as well as the setup—which includes how much power in each channel the BTS simulator is transmitting, the data rate, and any noise source being input (in this case, 4.5 dB of AWGN).

Testing the receiver of a mobile phone is similar to the methods used with TDMA systems discussed in Chapter 9. Again, the mobile can report the FER back to the test set or a loopback method can be used, where the test set would determine FER. Statistical analysis comes into play in this type of testing to ensure that the measurement is correct. This is the reason many times the specifications for a receiver test will be considered valid only if there is greater than a 95% confidence level.

Several methods are employed for testing the receiver of a BTS. The most common method is for the test set to transmit a reverse link signal with a known PN sequence. In essence, the BTS will make the measurement by estimating the FER. This measurement is then either reported back to the test instrument or to the BTS's local controller. Of course, the BTS and test set must have the same timing and frequency references for this to be accurate. Most test sets have the ability to transmit the proper signal with the pattern specified by standards. Another method that can be used to test a BTS receiver is to generate a reverse link signal and inject it into the receiver. Then, an IF is taken from the BTS and RHO is measured. As RHO should be known of the test signal, degradations of RHO would indicate a problem in the receive path.

Receiver tests include:

- *Sensitivity:* Sensitivity is the lowest level at which the receiver is able to produce a usable baseband. For the mobile and BTS, this requires measuring the FER while lowering the level below the specification.
- *Dynamic range:* It is important that the receivers are able to receive low levels, but it is also important that they are able to receive high-level signals as well. FER here is measured with the highest and lowest signals to ensure this dynamic range is sufficient.
- *Demodulation in AWGN:* AWGN simulates the noisy situations a mobile or BTS will see during operation. It is important that the receiver be able to receive under such situations. This involves using an AWGN generator and combining it with the signal being used to test the receiver. FER is then measured with this noise source connected.

13.10 1X and Beyond

As you saw in earlier chapters, there were some significant changes in CDMA with the introduction of lx and the high-speed data formats 1XEV-DO and 3X. Testing, however, remains very similar to what is described in this chapter. There will be some variations; however, if a technician has a good grasp of the testing described here, there should not be much issue learning the differences.

Testing data is not much different than testing voice—the RF metrics must be correct or the data transmission will lose quality, just as a voice transmission will. Throughput and data rate testing are often important measurements on data systems; however, due to the many protocol layers of the system, it is really as a go/no go test because a delay anywhere from RF to the application layer can cause degradations. Packet retransmissions (i.e., how many times a packet had to be re-sent) is often an excellent data measurement for RF technicians looking at network quality. Here problems in the RF will often manifest themselves, as RF is the primary cause of severe issues in this regard.

13.11 Conclusion

CDMA testing follows many of the same concepts used in other technologies. While power control testing is much more complex than other technologies, most of the other tests keep in line with common concepts. As technology in systems becomes more complex, test equipment must also evolve and become more complex. This then cancels out the complexity and allows technicians to focus on what they do best—maintaining systems—rather than learning the complexities of test engineering.

CHAPTER 14

WCDMA and Beyond: Testing Overview

14.1 Overview

After reading through the technology chapters, it should be clear to the reader that WCDMA is quite a complex format, with significant differences from 2G formats. It should also be understood that WCDMA is remarkably similar to CDMA2000. A common theme in Chapters 10 through 13 and this chapter is that the testing concepts remain very similar across the formats, while the specific methods will vary. This test is designed to just give a more generic overview of testing concepts. Thus, this chapter will not delve into too many specifics of WCDMA RF testing, as they are very similar to CDMA2000. Therefore, it is suggested that the reader looking to understand WCDMA testing concepts first read the CDMA testing chapter.

If a technician understands the differences in technology between CDMA2000 and WCDMA, understanding the differences in testing becomes rather simple.

14.2 Transmission Tests

Like CDMA, power accuracy it is very important in WCDMA to ensure proper network operation, because each code channel is transmitting on the same frequency. Thus, most of the standard WCDMA tests will relate to power accuracy. As mentioned earlier, there are no truly new tests, so most of these transmission tests will look familiar:

- *Maximum transmission power:* This ensures the BTS power amplifiers can transmit at their rated power level.
- *CPICH power accuracy:* It is imperative that the CPICH power is very accurate, as it will be used to set cell boundaries, and other channels will be related to it.
- *Frequency accuracy:* Ensuring the frequency is stable should be obvious.
- *Power control steps:* Because both the uplink and downlink have power control in WCDMA, it is important to make sure that the steps are accurate.

- *Power control dynamic range:* Naturally, a transmitter must be able to transmit across its whole range of power control, from lowest to highest. For instance, WCDMA base stations need to be able to transmit from their maximum output (tested earlier) to their lowest level (–28 dB).
- *Spectral emission:* Like any RF signal, it is important ensure that the signal is clean and is being filtered adequately so as not to cause interference to other frequencies. There are a variety of specific tests for this purpose, but the end concept is the same—ensure the signal gets transmitted where it is supposed to and not out of its assigned frequency.
- *Code-domain tests:* It is important to read through the CDMA testing chapter to understand code-domain tests. WCDMA code-domain tests are very similar. Specifically, a standard test of peak code domain error is used to ensure the code channels are accurate.
- *Modulation accuracy:* EVM is measured on the P-CCPCH channel.

14.3 Receiver Tests

Receiver tests in WCDMA will again sound similar to those in CDMA. Performing these tests involves demodulating to get a BER and then degrading the signal in various ways in order to ensure the receiver can keep the BER below a certain threshold. These tests would include:

- *Sensitivity:* This is the lowest level a signal can be received while maintaining a set BER.
- *Dynamic range:* This ensures signals can be received accurately across the level range of the receiver.
- *Adjacent channel selectivity:* This ensures the receiver can receive a signal adequately in the presence of another adjacent carrier.
- *Intermodulation characteristics:* This is similar to adjacent channel selectivity, except the interfering signal is created with two signals that create an intermodulation product.
- *Spurious emissions:* This simply ensures no spurious emanates from the receiver.

14.4 Conclusion

Obviously, this chapter is very short and general, primarily to avoid going in to details on tests that have already been discussed in detail in previous chapters. That said, as WCDMA is deployed further, testing techniques will mature as well and may change. Hence, if a technician has a good understanding of the technology, understanding how to test that technology will be substantially easier.

Appendix A

Watts to dBm

Watts	*dBm*
1,000,000	+90
100,000	+80
10,000	+70
1,000	+60
100	+50
10	+40
4	+36
2	+33
1	+30
0.5	+27
0.25	+24
0.1	+20
0.01	+10
0.001	0
0.0001	–10
0.00001	–20
0.000001	–30
0.0000001	–40
0.00000001	–50
0.000000001	–60
0.0000000001	–70
0.00000000001	–80
0.000000000001	–90
0.0000000000001	–100
0.00000000000001	–110

Appendix B

SWR, Reflection Coefficient, and Return Loss

SWR	*Reflection Coefficient*	*Return Loss (dB)*	*SWR*	*Reflection Coefficient*	*Return Loss (dB)*
17.391	0.8913	1	1.0580	0.0282	31
8.7242	0.7943	2	1.0515	0.0251	32
5.8480	0.7079	3	1.0458	0.0224	33
4.4194	0.6310	4	1.0407	0.0200	34
3.5698	0.5623	5	1.0362	0.0178	35
3.0095	0.5012	6	1.0322	0.0158	36
2.6146	0.4467	7	1.0287	0.0141	37
2.3229	0.3981	8	1.0255	0.0126	38
2.0999	0.3548	9	1.0227	0.0112	39
1.9250	0.3162	10	1.0202	0.0100	40
1.7849	0.2818	11	1.0180	0.0089	41
1.6709	0.2512	12	1.0160	0.0079	42
1.5769	0.2239	13	1.0143	0.0071	43
1.4985	0.1995	14	1.0127	0.0063	44
1.4326	0.1778	15	1.0113	0.0056	45
1.3767	0.1585	16	1.0101	0.0050	46
1.3290	0.1413	17	1.0090	0.0045	47
1.2880	0.1259	18	1.0080	0.0040	48
1.2528	0.1122	19	1.0071	0.0035	49
1.2222	0.1000	20	1.0063	0.0032	50
1.1957	0.0891	21	1.0057	0.0028	51
1.1726	0.0794	22	1.0050	0.0025	52
1.1524	0.0708	23	1.0045	0.0022	53
1.1347	0.0631	24	1.0040	0.0020	54
1.1192	0.0562	25	1.0036	0.0018	55
1.0935	0.0447	27	1.0028	0.0014	57
1.0829	0.0398	28	1.0025	0.0013	58
1.0736	0.0355	29	1.0022	0.0011	59
1.0653	0.0316	30	1.0020	0.0010	60

Glossary

1XRTT Abbreviation for first advanced phase of CDMA2000.

3GPP Third Generation Partnership Program. This is essentially a standards body looking at WCDMA.

3GPP2 Third Generation Partnership Program for CDMA2000.

3XRTT Abbreviation for three-carrier CDMA2000.

Access channel Generally, a reverse-link channel that a mobile station will use to gain access on the BTS.

Active channel A channel that is transmitting voice data specifically for a mobile station. In TDMA systems, there is one active channel at a time; in CDMA, there can be several.

Additive white Gaussian noise (AWGN) A noise source used for interference testing.

Adjacent channel interference Interference caused on one transmitted channel by another channel being transmitted at a nearby frequency.

Advanced Mobi1le Phone System (AMPS) The analog standard used extensively in North America as well as South America and parts of Asia.

Air interface Refers to the protocol used between base stations and mobile stations.

Alert An alert caused by a message from the BTS to the mobile station.

Algerbraic code excited linear prediction (ACELP) A vocoder technique used in both GSM and NA-TDMA.

American National Standards Institute (ANSI) The primary U.S. standards-approving body.

Associated control channel A channel that is associated with a particular voice channel. This generally relates to a logical data channel that uses the same physical channel as voice traffic.

Attenuator A device used to reduce the amount of power transmitted. Also called a pad, it is generally used at the front end of test instruments to prevent blowing out the instrument.

Authentication A process used to validate particular mobile stations.

Authentication center A database used to control the authentication process from the network side.

Authentication key (A key) The secret 64-bit key used in the North American authentication process, which is used to create shared secret data.

Backhaul The communications line between the base station and the BSC or MSC, be it fiber, microwave, or otherwise.

Bandwidth The specific range of frequencies that a signal can occupy.

Base transceiver system (BTS) The base station or cell site.

Base station controller (BSC) The part of a system that controls base stations and interfaces with the switching center (e.g., MSC in GSM).

Base station subsystem (BSS) Encompasses the base stations and BSC.

Bit Generally a binary signal element that directly represents data.

Bit-error rate (BER) The number of error bits related to the number of transmitted bits.

Blank and burst Interruption of user voice information in order to transmit important control information.

Block interleaving A method of scrambling transmitted bits.

Burst In TDMA systems, this generally refers to one time slot or transmission (in multislot operation, one burst may include several time slots; used for high-speed data transmission). In CDMA systems, this refers to one power control group.

Call processing The processes a mobile station or BTS will use to originate, accept, and maintain calls.

Carrier A signal frequency band. In TDMA and CDMA, this can carry several channels.

Customized applications for mobile enhanced logic (CAMEL) Refers to the enhanced intelligent network capabilities being added to GSM.

Candidate In CDMA, this refers to the PN offsets that a mobile station should monitor, as the BTS can decide that they should become active.

Cellular authentication and voice encryption algorithm (CAVE) Used in the authentication process to generate authentication results.

Cellular subscriber station (CSS) Another term for mobile station.

Cellular Telecommunications Industry Association (CTIA) An international organization that includes wireless manufacturers and carriers.

Chip A binary signal element in spread-spectrum systems that indicates that it is part of data that has been spread.

Closed-loop power control In CDMA, a power control system for mobile stations in which the BTS measures received power and orders the mobile to increase or decrease power accordingly.

Cochannel interference Interference caused on a frequency channel caused by another channel using the same frequency channel. This is usually the result of bad network planning.

Code-division multiple access (CDMA) The direct-sequence spread-spectrum system that separates users with unique codes.

Code domain Examining the properties (such as power, timing, and phase) of a CDMA signal in relation to each Walsh-encoded channel.

Common control channel A control channel that will carry information for all mobiles.

Constellation analysis A method of examining digital modulation graphically.

Control channel A logical or physical channel that carries control information to and from a mobile station.

Convolutional encoder The transmitter part of a forward correction system. Usually paired with a Viterbi decoder on the receive side.

Crosstalk When a conversation of one mobile is heard on another.

Cyclic redundancy check (CRC) An error-detecting process.

Dedicated control channel A control channel that is dedicated to a particular session.

Digital AMPS (D-AMPS) Another term for NA-TDMA.

Digital Cellular System 1800 (DCS 1800) GSM operating at 1,800 MHz.

Digital Cellular System 1900 (DCS 1900) GSM operating at 1,900 MHz. Also called PCS 1900.

Digital color code An identifier of base stations on the control channel.

Digital control channel (DCCH) A control channel that uses phase modulation.

Digital traffic channel (DTCH) The channel that carries digital voice conversations.

Discontinuous transmission A method of reducing a mobile station's transmit power when no speech activity is present.

Diversity Usually refers to the use of two antennas at different locations to receive the same signal to avoid fading. This is specifically called space diversity.

Downlink The forward channel, base station to mobile station.

Dual band A mobile station or BTS that uses more than one frequency band.

Dual mode A mobile station or BTS that uses more than one format (e.g., TDMA and AMPS).

Dual-tone multifrequency (DTMF) Tones used to represent digital bits over a voice channel.

Effective radiated power (ERP) The amount of power actually transmitted by an antenna.

Electronic serial number (ESN) A 32-bit number stored in the mobile station.

Enhanced Data for Global Evolution (EDGE) A high-speed packet-switched data format that allows for GSM operators to offer high-speed data service with a relatively easy evolutionary path.

Enhanced variable-rate vocoder (EVRC) An 8-kHz vocoder for CDMA that should provide better voice quality than the original QCELP 8-kHz vocoder.

Equalization The process of eliminating intersymbol interference.

Equipment identity register A database in the network that supplies mobile station information.

Erlang A measure of radio usage. One Erlang is equivalent to one telephone line being permanently used.

Error vector magnitude (EVM) A measurement of how accurate the phase modulation of a transmitter is.

Fading The degradation of the received signal due to a variety of reasons, including distance from the transmitter, multipaths, and shadowing.

Fast associated control channel (FACCH) A channel that overrides voice information to relay important control information during a voice conversation.

Federal Communications Commission (FCC) The regulating body of communication systems in the United States.

Flash A message sent on a voice channel usually causing an action. Typically, this message is sent when the user presses the SEND key during a conversation.

Forward channel From the BTS to the mobile station. This is also called the downlink.

Forward control channel (FOCC) In AMPS, the analog system control channel transmitted by the BTS.

Forward voice channel (FVC) The channel used for voice transmission on the forward link.

Frame-error rate (FER) The number of frame errors related to the number of transmitted frames.

Frequency-division duplex (FDD) The concept of using different frequencies for the forward and reverse channels.

Frequency-division multiple access (FDMA) Allowing multiple users simultaneous access by assigning each user a specific frequency channel.

Frequency hopping In GSM, this is the process of switching frequencies continuously in order to gain the benefits of spread-spectrum signals (e.g., a reduction of fading impact).

Frequency reuse The process of using the same frequency channels at different locations of a network, but planning enough propagation loss to avoid cochannel interference.

Gateway GPRS support node (GGSN) The data gateway from the GPRS PLMN to the outside data world.

Gaussian minimum shift keying (GMSK) The modulation technique used in GSM.

General packet radio service (GPRS) A packet data service being implemented in GSM and NA-TDMA that will allow much higher data rate service.

Global Positioning System (GPS) A satellite system that transmits extremely accurate time and synchronization signals and is also used for position locating.

Global System for Mobile Communications (GSM) The GSM system in use around the world. Originally, this was called Groupe Special Mobile.

Half rate Vocoders that use half the bandwidth of current vocoders and thus can double the amount of calls.

Handoff The process of transferring a mobile's call from one base station to another.

Handover Same as a handoff, but the GSM/European term.

Hard handoff A handoff in which the connection with the first serving base station must be cut before the connection to the next base station can begin. All NA-TDMA and GSM handoffs are hard handoffs.

High-speed circuit-switched data (HSCSD) Allows data service up to 64 Kbps over GSM.

Home location register (HLR) A database in the network that contains data about mobiles subscribed to the network.

Hybrid phase shift keying (HPSK) Complex modulation scheme used on CDMA2000 uplink to prevent zero crossings in the presence of multiple code channels.

Idle mode The mode a mobile station is in when not on a call and not getting access.

Intersystem handoff A handoff between two base stations that are connected to two different switches.

Intracell handoff A handoff between two frequency channels on the same base station.

Intermodulation Interference caused by mixing of certain frequencies.

International mobile equipment identity (IMEI) A serial number that can be used to identify stolen mobile stations.

International mobile subscriber identification (IMSI) A telephone number that conforms to worldwide standards.

Intersymbol interference When the same signal takes two paths to a receiver (a multipath), one will arrive sooner than the other—the same bit will arrive at different times and cause interference.

Linear prediction coding with regular pulse excitation (LPC-RPE) A vocoding technique used in GSM.

Logical channel Regardless of the method of transmitting, a logical channel simply refers to a channel that carries ones and zeros. It may be its own physical channel or it may be associated into an existing channel.

Long code A code sequence used in CDMA that has an absolute sequence position tied to the current time.

Mobile assisted handoff (MAHO) The process of using the mobile station to perform measurements in order to facilitate the handoff process.

Modulation efficiency The number of transmitted bits per second for each Hertz used.

Mobile station integrated services digital number (MSISDN) The number dialed to call a GSM mobile station.

Mobile identity number (MIN) The telephone number of a mobile station.

Multipath propagation When a signal takes more than one path to reach a receiver. See also intersymbol interference.

Mobile switching center (MSC) The connection between the mobile system and the public phone system.

Mobile telephone switching office (MTSO) The mobile switch used in AMPS to control base stations and connect to the public switch.

Narrowband AMPS (N-AMPS) A derivative of AMPS developed by Motorola to increase the capacity of the analog system.

Neighbor list In CDMA, a list of PN offsets of the nearby base stations.

Nordic Mobile Telephone (NMT) An early analog system originally launched in Scandinavia.

North American TDMA (NA-TDMA) The term used by this text for the TDMA system referred to in the IS-136 standard.

Origination The process of beginning a phone call. This typically refers to a mobile origination, which means the mobile is originating the call.

Orthogonal sequence Sequences that are completely uncorrelated to each other. In CDMA, it refers to the Walsh sequences used to separate users.

Open loop The power control subsystem in CDMA that makes an early estimate of transmit power based primarily on received power.

Page The process of a BTS-originated call.

Paging channel In CDMA and GSM, the channel a mobile station will monitor to see if a call is coming in.

Personal Communications System (PCS) Usually refers to the 1,900-MHz frequency band; although more and more, it simply refers to mobile communications systems with high-end features, regardless of the band they utilize.

Pilot channel In CDMA, a channel used for initial timing and BTS location.

PN offset In CDMA, the offset on the BTS transmitted signal that identifies particular base stations.

PN pollution A network planning problem in CDMA, where too many base station signals are reaching one area, causing the noise floor to raise.

Processing gain The ratio between the transmit rate of a signal and the actual effective bit rate. Only applies to spread-spectrum systems.

Public land mobile network (PLMN) The wired side of a communications network.

Public switched telephone network (PSTN) The public switch a mobile network must interface to in order to make calls to the outside world.

Pseudorandom (PN) sequence A type of sequence that has the characteristics of a random sequence, but is, in fact, known.

Qualcomm code excited linear prediction (QCELP) The vocoder technique used in CDMA.

Quasi-orthogonal Walsh codes (QOF) Extra orthogonal codes created by applying a mask to Walsh codes.

Rake receiver A receiver used in CDMA that takes advantage of multipath signals by coherently combining them.

Random access channel A channel used by mobiles to access the system.

Received signal strength indicator (RSSI) An indicator from a mobile to a BTS informing the network of the received signal strength.

Registration The process of a mobile giving its information to the network.

Release The process of a mobile station or base station ending a call.

Repeater A device used to extend the range of a cell by receiving a signal and rebroadcasting the same signal on a different frequency.

Return loss A measurement of the amount of signal energy that does not make it out of a transmission path (usually an antenna). This is relative to the total transmitted signal.

Reverse control channel (RECC) The channel used by a mobile station to access a system.

Reverse voice channel (RVC) The channel used for voice traffic on the reverse link.

Rho Also called waveform quality. In CDMA, this is a correlation measurement that indicates transmitter performance.

Roaming The process of using a mobile station in a different network than it was subscribed in.

Serving GPRS support node (SSGN) The gateway to the mobiles from the GPRS PLMN.

Sectorization The process of dividing a BTS into pie-shaped cells using directional antennas.

Shared channel feedback (SCF) Used on reverse channel access to let a mobile know when to try to access.

Shared secret data (SSD) A code created using the A key that will be input into the CAVE algorithm in order to create authentication results.

Short message service (SMS) The ability to send messages up to 160 characters to and from mobile stations.

Signal-to-noise (S/N) ratio A ratio of the carrier power to the background noise of a channel.

SINAD The basis for receiver tests in analog systems.

Sleep mode The process of a mobile only monitoring the paging channels when it needs to in order to conserve battery power when not on a call.

Slow associated control channel (SACCH) An associated control channel within the traffic channel that relays pertinent, but not critical, information to the mobile station.

Soft handoff In CDMA, the ability of a mobile station to be able to receive more than one BTS signal at a time, so that it can make a connection to a new BTS before it breaks the connection to the old BTS.

Softer handoff A soft handoff between two sectors of a BTS.

Standalone dedicated control channel (SDCCH) In GSM, a channel dedicated to a particular mobile station for control channel operation.

Station class mark (SCM) An identifier of the mobile station that indicates the mobile's capabilities.

Subscriber identity module (SIM) A card used in GSM that carries all of the pertinent subscriber information; it can be transported to different GSM phones.

Supervisory audio tone (SAT) One of three tones used to identify particular base stations.

Supervisory tone A 10-kHz tone used by the mobile to signal certain events.

Synchronization channel (SYNC) A channel used to get the mobile station synchronized with the BTS.

System identifier (SID) An identifying code sent by base stations indicating a particular operator and geographic area.

System operator code (SOC) A code used to identify a particular operator.

Telecommunications Industry Association (TIA) An association that organizes standards bodies for the telecommunications industry.

Temporary mobile subscriber identity (TMSI) A temporary identifier assigned by the network to a mobile station.

Time-division duplex (TDD) The concept of using different time slots for forward and reverse channels. It is not used in the high-tier wireless formats discussed in this book, but it is used in cordless formats such as DECT and PHS.

Time-division multiple access (TDMA) The process of multiplexing multiple users on a single frequency channel using different time slots. It is used in NA-TDMA and GSM.

Time slot A time allotted for a particular user.

Timing advance The function of a mobile to change its timing to account for propagation delay.

Total Access Communications System (TACS) An early analog system used in Europe.

Universal Mobile Telecommunications Service (UMTS) A name given to a third generation wireless system, typically the European entry.

User equipment (UE) WCDMA name for a mobile device.

Uplink Another term for reverse link.

Vector sum excitation linear prediction (VSELP) The original vocoder technique used in NA-TDMA.

Viterbi decoder The receiver pair of a convolutional encoder.

Visitor location register (VLR) A database that holds information on mobiles visiting a network.

Voltage standing wave ratio Relates to return loss, an indicator of antenna and transmission path operation.

Walsh codes A family of orthogonal sequences used in CDMA.

Wireless local loop (WLL) The use of wireless transmissions to replace copper wiring for connecting the subscriber to the telephone network.

Selected Bibliography

Bates, R. J., *GPRS,* New York: McGraw-Hill, 2002.

Foster, W., and S. Manning, *GSM Introduction,* Ismanning, Germany: Wavetek GmbH, 1998.

Gallagher, M. D., and R. A. Snyder, *Mobile Telecommunications Networking with IS-41*, New York: McGraw-Hill, 1997.

Garg, V. K., K. F. Smolik, and J. E. Wilkes, *Applications of CDMA in Wireless/Personal Communications,* Upper Saddle River, NJ: Prentice Hall, 1997.

Goodman, D. J., *Wireless Personal Communications Systems*, Reading, MA: Addison-Wesley, 1997.

Harte, L., R. Levine, and R. Kikta, *3G Wireless Demystified,* New York: McGraw-Hill, 2002.

Kahabka, M., *Pocket Guide for Fundamentals and GSM Testing*, Eningen, Germany: Wandel & Goltermann GmbH & Co, 1998.

Korhonen, J., *Introduction to 3G Mobile Communications,* Norwood, MA: Artech House, 2001.

Lee, W. C. Y., *Mobile Cellular Telecommunications*, New York: McGraw-Hill, 1995.

Steele, R., C. Lee, and P. Gould, *GSM, cdmaOne, and 3G Systems,* New York: John Wiley & Sons, 2001.

Webb, W., *Understanding Cellular Radio*, Norwood, MA: Artech House, 1998.

About the Author

Andrew Miceli has been working in the communications industry since 1991, being introduced to wireless communications while serving in the U.S. Army. He received his B.S. from LeTourneau University in Longview, Texas. Mr. Miceli has held various positions with test and measurement organizations and was involved primarily with cellular base station and mobile station test instruments until 2000, when he joined Superconductor Technologies, Inc. He began his career as a repair technician at Anritsu Corporation, and he has held application engineering, sales support engineering, sales support management, and strategic marketing management positions with Wavetek Corporation. Presently, Mr. Miceli serves as the regional sales vice president for Superconductor Technologies, selling superconductor-based filters and components to wireless operators.

Over the last several years, Mr. Miceli has written numerous training programs that have been used to educate hundreds of wireless technicians around the world. He has written numerous articles for industry magazines and has presented technical presentations at several industry events. His primary focus in recent years has been on network-enhancement applications, working closely with wireless operators to improve the quality of wireless.

Index

D

G

H

U

V

Recent Titles in the Artech House Mobile Communications Series

John Walker, Series Editor

Advances in 3G Enhanced Technologies for Wireless Communications, Jiangzhou Wang and Tung-Sang Ng, editors

Advances in Mobile Information Systems, John Walker, editor

CDMA for Wireless Personal Communications, Ramjee Prasad

CDMA Mobile Radio Design, John B. Groe and Lawrence E. Larson

CDMA RF System Engineering, Samuel C. Yang

CDMA Systems Engineering Handbook, Jhong S. Lee and Leonard E. Miller

Cell Planning for Wireless Communications, Manuel F. Cátedra and Jesús Pérez-Arriaga

Cellular Communications: Worldwide Market Development, Garry A. Garrard

Cellular Mobile Systems Engineering, Saleh Faruque

The Complete Wireless Communications Professional: A Guide for Engineers and Managers, William Webb

EDGE for Mobile Internet, Emmanuel Seurre, Patrick Savelli, and Pierre-Jean Pietri

Emerging Public Safety Wireless Communication Systems, Robert I. Desourdis, Jr., et al.

The Future of Wireless Communications, William Webb

GPRS for Mobile Internet, Emmanuel Seurre, Patrick Savelli, and Pierre-Jean Pietri

GPRS: Gateway to Third Generation Mobile Networks, Gunnar Heine and Holger Sagkob

GSM and Personal Communications Handbook, Siegmund M. Redl, Matthias K. Weber, and Malcolm W. Oliphant

GSM Networks: Protocols, Terminology, and Implementation, Gunnar Heine

GSM System Engineering, Asha Mehrotra

Handbook of Land-Mobile Radio System Coverage, Garry C. Hess

Handbook of Mobile Radio Networks, Sami Tabbane

High-Speed Wireless ATM and LANs, Benny Bing

Interference Analysis and Reduction for Wireless Systems, Peter Stavroulakis

Introduction to 3G Mobile Communications, Second Edition, Juha Korhonen

Introduction to GPS: The Global Positioning System, Ahmed El-Rabbany

An Introduction to GSM, Siegmund M. Redl, Matthias K. Weber, and Malcolm W. Oliphant

Introduction to Mobile Communications Engineering, José M. Hernando and F. Pérez-Fontán

Introduction to Radio Propagation for Fixed and Mobile Communications, John Doble

Introduction to Wireless Local Loop, Second Edition: Broadband and Narrowband Systems, William Webb

IS-136 TDMA Technology, Economics, and Services, Lawrence Harte, Adrian Smith, and Charles A. Jacobs

Location Management and Routing in Mobile Wireless Networks, Amitava Mukherjee, Somprakash Bandyopadhyay, and Debashis Saha

Mobile Data Communications Systems, Peter Wong and David Britland

Mobile IP Technology for M-Business, Mark Norris

Mobile Satellite Communications, Shingo Ohmori, Hiromitsu Wakana, and Seiichiro Kawase

Mobile Telecommunications Standards: GSM, UMTS, TETRA, and ERMES, Rudi Bekkers

Mobile Telecommunications: Standards, Regulation, and Applications, Rudi Bekkers and Jan Smits

Multiantenna Digital Radio Transmission, Massimiliano "Max" Martone

Multipath Phenomena in Cellular Networks, Nathan Blaunstein and Jørgen Bach Andersen

Multiuser Detection in CDMA Mobile Terminals, Piero Castoldi

Personal Wireless Communication with DECT and PWT, John Phillips and Gerard Mac Namee

Practical Wireless Data Modem Design, Jonathon Y. C. Cheah

Prime Codes with Applications to CDMA Optical and Wireless Networks, Guu-Chang Yang and Wing C. Kwong

QoS in Integrated 3G Networks, Robert Lloyd-Evans

Radio Engineering for Wireless Communication and Sensor Applications, Antti V. Räisänen and Arto Lehto

Radio Propagation in Cellular Networks, Nathan Blaunstein

Radio Resource Management for Wireless Networks, Jens Zander and Seong-Lyun Kim

RDS: The Radio Data System, Dietmar Kopitz and Bev Marks

Resource Allocation in Hierarchical Cellular Systems, Lauro Ortigoza-Guerrero and A. Hamid Aghvami

RF and Microwave Circuit Design for Wireless Communications, Lawrence E. Larson, editor

Sample Rate Conversion in Software Configurable Radios, Tim Hentschel

Signal Processing Applications in CDMA Communications, Hui Liu

Software Defined Radio for 3G, Paul Burns

Spread Spectrum CDMA Systems for Wireless Communications, Savo G. Glisic and Branka Vucetic

Third Generation Wireless Systems, Volume 1: Post-Shannon Signal Architectures, George M. Calhoun

Traffic Analysis and Design of Wireless IP Networks, Toni Janevski

Transmission Systems Design Handbook for Wireless Networks, Harvey Lehpamer

UMTS and Mobile Computing, Alexander Joseph Huber and Josef Franz Huber

Understanding Cellular Radio, William Webb

Understanding Digital PCS: The TDMA Standard, Cameron Kelly Coursey

Understanding GPS: Principles and Applications, Elliott D. Kaplan, editor

Understanding WAP: Wireless Applications, Devices, and Services, Marcel van der Heijden and Marcus Taylor, editors

Universal Wireless Personal Communications, Ramjee Prasad

WCDMA: Towards IP Mobility and Mobile Internet, Tero Ojanperä and Ramjee Prasad, editors

Wireless Communications in Developing Countries: Cellular and Satellite Systems, Rachael E. Schwartz

Wireless Intelligent Networking, Gerry Christensen, Paul G. Florack, and Robert Duncan

Wireless LAN Standards and Applications, Asunción Santamaría and Francisco J. López-Hernández, editors

Wireless Technician's Handbook, Second Edition, Andrew Miceli

For further information on these and other Artech House titles,
including previously considered out-of-print books now available through our
In-Print-Forever® (IPF®) program, contact:

Artech House
685 Canton Street
Norwood, MA 02062
Phone: 781-769-9750
Fax: 781-769-6334
e-mail: artech@artechhouse.com

Artech House
46 Gillingham Street
London SW1V 1AH UK
Phone: +44 (0)20 7596-8750
Fax: +44 (0)20 7630-0166
e-mail: artech-uk@artechhouse.com

Find us on the World Wide Web at:
www.artechhouse.com